Physics in Mind

Physics in Mind

A Quantum View of the Brain

Werner R. Loewenstein

BASIC BOOKS

New York
A Member of the Perseus Books Group

Books published by Basic Books are available at special discounts for bulk purchases in the United States by corporations, institutions, and other organizations. For more information, please contact the Special Markets Department at the Perseus Books Group, 2300 Chestnut Street, Suite 200, Philadelphia, PA 19103, or call (800) 810-4145, ext. 5000, or e-mail special.markets@perseusbooks.com.

Designed by Trish Wilkinson
Set in 11.5 point Minion Pro

Library of Congress Cataloging-in-Publication Data

Loewenstein, Werner R.
    Physics in mind : a quantum view of the brain / Werner R. Loewenstein.
        p.   cm
    Includes bibliographical references and index.
    ISBN 978-0-465-02984-6 (hardcover) — ISBN 978-0-465-03397-3 (e-book)
    1. Neurophysiology. 2. Neural networks (Neurobiology) 3. Brain. I. Title.
QP355.2.L64 2013
612.8—dc23   2012028268

10 9 8 7 6 5 4 3 2 1

To my children, Claudia, Patricia, Harriet, and Stewart

Contents

Preface   *xiii*

A Thumbnail Sketch of Our Journey   *xvii*

1 Our Sense of Time: Time's Arrow   1

    Our Awareness of Time   1

    The Stream of Consciousness   2

    Inner Time versus Physics Time   4

    The Way the Cookie Crumbles   6

    How to Predict the Future from the Past   8

    Why the Cookie Crumbles   11

    Can Time Go Backward?   12

    Time and Our Perception of Reality   13

2 Information Arrows   15

    Retrieving the Past   15

    The Arrows of Time and Information   19

    The Products of Information Arrows: An Overview   21

    Life's Arrow   22

    The Strange Circles at Coordinates 0,0,0   25

    Molecular Demons   26

    Wresting Information from Entropy:

        The Quintessence of Cognition   28

    Who Shoulders Life's Arrow?   32

3 The Second Coming   35

    The Demons for Fast Information Transmission   36

    The Sensory Demons   41

A Generalized Sensory Scheme   42
The Demon Tandem   44
How the Demon Tandems Censor Incoming Information   46

4 The Sensors   49

The Sensory Transducer Unit   49
A Lesson in Economics from a Master   51
The Silent Partner   53
How Electrical Sensory Signals Are Generated   54

5 Quantum Sensing   57

The Quantum World   58
Our Windows to the Quantum World   61
Coherent Quantum Information Transmission   64
The Advantages of Being Stable   66
Why We See the Rainbow   67
The Demons Behind Our Pictures in the Mind:
    A Darwinistic Physics View   70
Why White Is White   72
The Quantum View   75
Again, Why White Is White   76
Lady Evolution's Quantum Game Plan   77
Quantum Particles That Don't Cut the Mustard   78

6 Quantum into Molecular Information   81

Boosting the Quantum   81
A Consummate Sleight of Hand   83
The Ubiquitous Membrane Demon   86

7 Molecular Sensing   87

A Direct Line from Nose to Cortex   87
A Thousand Information Channels of Smell   89
Mapping, Coding, and Synonymity   90
Molecular Sensory Synonymity   92
Why Sensory Synonymity   95
Quantum Synonymity   97
Harmless Double Entendres   98

8 Electronic Transmission of Biological Information 99

Evolution's Favorite Leptons 99
Electronic Information Transmission:
   A Development Stumped 101
Two Old Batteries 103

9 The Random Generators of Biomolecular Complexity 107

Genuine Transmogrification 107
A Quantum Random Generator of Molecular Form 109
The Heuristics of Transmogrification 113
The Random Generator and Our Genetic Heritage 114
An Algorithm Is No Substitute for a Demon 116
The Second Generator of Biomolecular Form 118
Complexity as a Windfall 122
Ikats 122

10 The Ascent of the Digital Demons 125

Quantum Electron Tunneling 125
The Electronic Cul-de-Sac 127
The Rise of the Digital Demons 127
Do Plants Have Digital Demons, Too? 128

11 The Second Information Arrow and
Its Astonishing Dénouement: Consciousness 131

The Structure of Time 132
The Evolutionary Niche in the Structure of Time:
   A Hypothesis 133
Forecognition 134

12 How to Represent the World 137

The Universal Turing Machine 137
Rendering the World by Computer 140
The Neuronal Virtual-Reality Generator 142
Our Biased World Picture 143
Computing by Neurons 145
Correcting Our World Picture 148
Flying the Coop of Our Senses 149

13 Expanded Reality   155

A Fine Bouquet   155
Mathematics and Reality   156
The Neuron Circuitry of Language   157
The Feathers of the Brain   159
Mathematics and Forecognition   160
The Reluctant Sensory Brain   161
The Limits of Knowledge   162
A Note About Reality   163

14 Information Processing in the Brain   165

Cell Organization in the Brain   166
Cortical Information-Processing Units   168
Cortical-Cell Topography and Worldview   170
A Last-Minute Change in Worldview   173
Retrieving a Lost Dimension   175
Information Processing in the Brain from the Bottom Up   178
Being of One Mind   179
Two Minds in One Body?   182
An Old Pathway Between Brain Hemispheres for
    Information Producing Emotion   183
The Virtues of Parallel Computation   184

**15 Information Transforms in the Cortex
and the Genesis of Meaning**   189

The Censoring of Sensory Information   189
What the Eyes Tell the Brain   191
Coding for the Vertical, the Horizontal,
    and the Oblique   193
The Neuron Pecking Order   194
The Grandmother Cell   196
Distributed Coding   200
The Imprinting of Cortical Cells   202
The Begetting of Meaning   204
How to Measure Meaning   205
Logical Depth   206
Lowbrow Meaning   208

16 The Conscious Experience   215

    The Consciousness Polychrome   216
    Consciousness, Sensory Information Processing,
        and Computing   216
    The Conscious Experience of the Passing of Time,
        and Memory   217
    Unconscious Thinking   219
    From Piecemeal Information to Unity of Mind   221
    Feelings and Emotions   223
    Gut Feelings   226
    Two Ways for Information to Get to the Promised Land 227
    The Virtues of Signal Synchrony as a Higher-Order
        Sensory Code   228
    Signal Synchrony and Conscious Perception   229
    Neuronal Synchronization during Competition for
        Access to Consciousness   231

17 Consciousness and Quantum Information   235

    Boltzmann's World   235
    Where the Quantum World May Come into
        Evolution's Ultimate Information Enterprise 237
    Quantum Information Waves   238
    The Vanishing Act   240
    Are Quantum Waves and Consciousness Entangled?   242
    Irretrievable Quantum Information Losses: Decoherence 243
    Why Our Weltanschauung Is So Narrow:
        An Unexpected Lesson for Philosophy   245
    On the Possibility of Quantum Coherence in the Brain 246

18 Molecular Quantum Information
   Processing and Quantum Computing   249

    Quantum Computers   250
    The Advantages of Quantum Computing   251
    Quantum Computers in the Real: Computing with Atoms 252
    Quantum Computing with Atomic Nuclei   254
    Natural Quantum-Computer Games   256
    How to Rig a Molecular Quantum Computer   259
    Toward Multi-Qubit Computers   261

19 Quantum Information Processing and the Brain   265

    Naturally Slow Decoherence   265
    A View of Evolution from the Quantum Bottom   266
    Two Variations on an Ancient Biological Quantum Theme   267
    A Final Darwinistic-Physics Inquirendo and a Hypothesis   269
    Neuronal Quantum Computing in Light of the Hypothesis   271
    Explanatory Remarks on the Hypothesis   273
    Prospects   274

Appendix   *277*
Recommended Reading   *281*
References   *283*
Acknowledgments   *309*
Illustrations and Other Credits   *311*
Index   *313*

Preface

The radio was on in the background when I became aware of some vaguely familiar chords. As I began to listen more intently, a few swells of harmony pushed a button somewhere: Bach, the chorale "Out of the Deep," the faint final accords.

I hadn't heard that piece in decades. But as a teenager I had sung it in the school choir. Hearing those notes, I remembered the music and words of the entire chorale, as well as the withering look of the conductor when I missed my cue . . . and yes, the face of that girl in the choir I had fancied.

A melody, a face, the sounds of rushing wind, the smell of honeysuckle, the touch of a hand long still—all this we can perceive with the mind's eye. We see, we hear, we feel, we remember, we are aware.

But what precisely do we mean when we say, "We are aware of something"? What is this peculiar state, at once so utterly familiar and so bewilderingly mysterious, that we call consciousness? What is its *mechanism*?

I put it like that point blank, to show from the start the tenor of the way and hold implicitly forth the expectation that consciousness has a physics explanation. Such a prospect may be shocking to some. That our mind and perceptions, our joys and sorrows, our memories, our sense of self, or worse, the glittering jewel of human intellect, thought, could be reduced to physics terms, may be a blow to one's self-esteem. But it is really no more so than anything evolutionary—Darwinian schemes always step on the peacock tail.

In any case, that prospect should no longer be as shocking as it might have been, say, 20 years ago. In the meantime, the neurosciences have advanced on a broad front, only to bolster reductionist aspirations. The advances have held up the mirror to the brain, its intricate web of a trillion neurons, letting us see in detail as never before the stream of information that nurses our perceptions and the information processing that precedes them.

I wrote this book in an attempt to make these advances accessible to a wide spectrum of readers. I center on the information processing that takes place at the sensory periphery of the brain and at the brain cortex and examine it in the light of information theory. I have been fortunate that in the past few years there has been a major breakthrough in an offspring of that theory, quantum computation; the most spectacular advances happened just as I penned the last three chapters. So I was able to view the sensory information processing in this new light, especially the parallel processing that is the hallmark of the cognitive brain. That processing is the antecedent of consciousness and is exquisitely sensitive and fast, offering a target to test one's reductionist mettle.

I originally intended to limit my story to the brain's sensory periphery, to the capture and transfer of information at sensory receptors, a field I had worked in early in my career. But as I went on with the story, my leading characters, a set of talented biological molecules, started to develop wills of their own and did things quite different from those I had planned for them. Well, I should have been forewarned. Those are molecules operating by the strange rules of quantum logic. And in no time they took over the brain's ground floor, the quantum bottom, presenting a tableau in which the boundaries between biology and physics vanished into thin air. That was a sight I could not resist. So this became a book on the sensory brain outright.

The book is written for the general reader with an interest in science. No specialized knowledge of biology, physics, or information theory is assumed in advance. With the general reader in mind, I have dispensed with mathematical apparatus. The few equations I used are tucked away in a footnote and the appendix. But the reader who wishes to skip them will not lose the thread of the book; the concepts they embody are explained in plain language along the way.

I have taken the liberty to personify throughout the book the process of biological evolution. I hope the reader won't mind. Such a personification comes naturally if one looks at things from the information angle. It simplifies the narrative of an evolutionary process that is rare, if not unique, in the physics universe, where the good throws of the dice needn't be repeated over and over again—an evolutionary process that generates its own information repository to progressively reduce the element of chance.

It may be surprising that in a book on brain and mind there should appear more physicists than biologists. This merely reflects the fact that the mind is frontier territory. Indeed, it is not at all uncommon in biological history to find physical scientists at the leading edges. Even Darwin, contrary to popular belief, was originally not a biologist, or at least he didn't think of himself as one when he set out on the voyage of the *Beagle*: "I a geologist have illdefined [sic] notion of land covered with ocean, former animals, slow force cracking surface . . . ," he wrote in his notebook. Nor did things change very much in that regard a century later, when modern biology was well on its way and molecular genetics was still a territory of uncharted wonders. Then again physicists were among the pioneers. And an encore is happening these days as the mind is becoming the new frontier. Indeed, the brain-mind problem, a subject that for centuries had been lying uneasily at the border of science and philosophy, may be the natural meeting ground between biologists and physicists.

Some years ago a group of students and colleagues of mine at Columbia University staged a mock biophysics symposium on my birthday. The "symposium" was an elaborate spoof on the vagaries of biophysics and the crowings of its practitioners. It was great fun, though much of it I have forgotten. But not the refrain of one of the ditties, and it has been haunting me ever since: "Biology is biology, and physics is physics, and never the twain shall meet." If this book helps to sink that refrain into oblivion, that is the best I would ask for.

A Thumbnail Sketch of Our Journey

After a brief excursion into our sense of time, our journey will take us to the origin of all information in the universe, and we will see what that gossamer called Information really is and how it got to where it is. Above all, we'll see the steep price sentient beings must pay for acquiring it and the natural limits to cognition. With that knowledge in the bag, we take a journey along the sensory information stream of the brain— the stream that nurses our cognitions. There, one by one, we will meet the neuronal entities who drive the stream, the prime movers who wrest information from the quantum and the molecular world outside, and those higher up in the brain who gradually extract meaning from the information hodgepodge in the stream, ending up with a coherent representation of the world outside.

CHAPTER ONE

Our Sense of Time: Time's Arrow

Our Awareness of Time

We begin our journey with an exploration of our sense of time. This sense is a prominent feature of consciousness, and I mean here not the sensing of periodicities and rhythms (the causes of which are reasonably well understood), but something more fundamental: the sensing of time itself, the passing of time. This we feel as something intensely real, as a constant streaming, as if there was an arrow inside us pointing from the past to the future.

This arrow is an integral part of our conscious experience, and more than that, a defining part of our inner selves. Yet for all its intimacy and universality, it has defied scientific explanation. Not that this is the only perverseness of consciousness, but I single it out because our sense of time appears to be the least ethereal and with some prodding, it might give ground. It has a measurable counterpart in physics, the quantity t. Indeed, that t occupies a high place in our descriptions of the physical universe—it enters into all equations dealing with events that evolve. However, we should be forewarned that that time and the one we sense may not be the same; t is a sort of housebroken variety of time, a variety that was tamed through mathematics. Nevertheless, it still has some wild qualities and, for something that has gone so thoroughly through the wringer of mathematics, it is surprisingly human.

1

This could hardly have been said before the revolution Einstein wrought. The *t* then was something that flowed uniformly and inexorably at the same universal rate, whether there was a human being there to observe it or not. It was, and I use Newton's own words, an "absolute, true and mathematical time, [which] of itself, and from its own nature, flows equably without relation to anything external." With a time like that, it didn't matter *who* measured it, as long as a good clock was used. The theory of relativity put an end to this disemboweled, machinelike time. In contemporary physics each observer must take his or her own measure of it. Clocks, however good, do not necessarily mark the same time in the hands of different observers. Far from flowing equally, time varies, depending on the relative movement of the observer. This kind of notion is something one can warm to, though it may not have all the tints and hues of our inner time.

The Stream of Consciousness

But first, what is this thing we call time? St. Augustine, one of the early inquirers and among the most profound, captured the zeitgeist when he wrote at the turn of the fifth century in his *Confessions*, "If no one asks me, I know; but if someone wants me to explain it [time], I don't know." He might as well have said it today. We are still groping, and if we ponder the question in terms of current physics, we may well wonder why time should have a direction at all.

The roots of our sense of time are somehow interwoven with those of our conscious states. They can be traced to the transience of these states—their constant flitting, one state following another. Ordinarily those states are so many and follow each other so fast that one doesn't see the forest for the trees. But that gets better if one fixes on a particular sequence—one streamlet in the stream. Try this, for example: have someone tap the tip of your index finger rhythmically with a pencil, spacing the taps half a second apart. You will feel a series of touch sensations that jointly convey a forward sense of time. The nerves in your finger send the tactile information in digital form to your brain; it takes somewhat under 50 milliseconds for that information to get to the cortex, and some 200–300 milliseconds for it to be processed at the various stations of the brain web and become a conscious state—enough time to give rise to an unslurred sequence of cognitive events. Now change the condition slightly and tag the sequence by making the second tap stronger than the first. You will feel the sensations in a distinct order, an outcome as revealing as it is plain: the conscious states are well ordered, and the order reflects that of the peripheral stimuli.

The same can be said for the states underlying our auditory and visual experiences. A series of notes played on the piano is heard in the sequence the notes were played, and a series of pictures flashed on a TV screen is seen in the same serial sequence. There are limits to the speed with which our nervous system can handle and resolve individual events, but *within these limits, our conscious states always occur in an unjumbled sequence.**

This also holds true for multiple sensory input. We see the finger movements of a pianist in the same sequence as the notes she produces, or we hear the taps of the tap dancer in step with his movements. Not even an illusion—such as hearing an extra note when there was none or seeing a phantom dance step—will jumble the picture. The bogus information is merely inserted into the sequence. And when memory comes into play, the imagined information reels off with the true information.

I bring up the matter of bogus images because it bears on the conscious states originating from within our brain in the absence of an immediate sensory input. Such intrinsic states seem to occur in orderly sequence, too—at least those in logical reasoning do. It is not by chance that we speak of a "train" of thought. Moreover, when we try to transmit this sort of information to somebody else, we tend to do so step by step, in a concatenated series. Indeed, all our language communications—be they in English, Chinese, mathematical symbols, clicks, or grunts—are based on systematic, ordered information sequences.

Thus a good part of our consciousness seems to be based on an orderly streaming of information states—a stream broken only during

---

*This rule applies even to speeds somewhat above these limits, though the events then get slurred. Indeed, this is what enables us to watch TV or see movies in which the individual pictures, the frames, last fewer than 45 milliseconds.

our sleeping hours, and sometimes not even then. It is probably this constant streaming that gives us a sense of the flow of time.

Inner Time versus Physics Time

So much for the flow. But what of its direction? Why the asymmetry of time, the never-deviating progression from past to future? Offhand, there seems to be no intrinsic reason why that flow should always have the direction it has. Indeed, from the point of view of physics, it might as well go the other way round.

Take Newtonian physics, for example. There the time t describing the evolution of a system has no preferred direction—it flows forward, as in our stream of consciousness, but it can also flow backward. The laws governing clockwork going in reverse are the same as those of clockwork going the usual way—the variable merely becomes $-t$. Time in Newtonian physics is inherently symmetrical, and no consideration of boundary condition has any significant bearing on this pervasive property.

The same may be said for all of physics. All the successful equations—from Hamilton's to Maxwell's to Einstein's to Schrödinger's—can be used as well in one time direction as in the other. In Einstein's relativity, time doesn't even flow. It is interwoven with space into one fabric, *space-time*. This fabric is a coordinate system, something static, so time flows here no more than space does. And if we artificially reverse the way things normally unfold, say by running a film of planets and stars backward, their movement still would conform to Einstein's laws.

Such symmetry also holds for quantum physics, the physics of small-scale things (10^{-14} cm and less). Although the theories here made a clean sweep of many traditional physics concepts, including some aspects of the notion of time, they held fast to the past in regard to the symmetry of time. They took over Newton's notion of an absolute and universal time, and just as a Newtonian system evolving in time traces a trajectory in space, a quantum system traces it in an infinite-dimensional state space. Thus in the micro-domain, the time dimension is not a mere Einsteinian coordinate, but rather something that flows, and it does so with no asymmetries.* As in the macro-domain, there is no fundamental distinction between past and future—the future merely repeats the past.

All this is a far cry from the way we perceive time in our conscious experience, and it makes one wonder why we feel time as something always progressing in one direction. Well, perhaps the Queen had it right when she explained to Alice the run of things in Wonderland:

Alice: "I don't understand you. It's dreadfully
         confusing!"

Queen: "That's the effect of living backwards,
         it always makes one a little giddy at first -- "

Alice: "Living backwards? I never heard of such a
         thing!"

Queen: " -- but there's one great advantage in it,
         that one's memory works both ways."

Alice: "I'm sure mine only works one way,
         I can't remember things before they happen."

Queen: "It's a poor sort of memory that only works
         backwards."

---

*The only exceptions are black holes and Anderson-Higgs asymmetries. In black holes, owing to their enormous gravity fields, the spacetime curvature becomes so large that it pulls the spacetime fabric apart. Time then ceases to exist altogether and the only thing that remains of the Einsteinian fabric is a probabilistic space. The Anderson-Higgs asymmetries are very small and unrelated to the ever-prevailing asymmetry, time's arrow, in the macro-domain.

The Way the Cookie Crumbles

Yet we have every reason to believe that the distinction between past and future, the time asymmetry, reflects reality—well, at least the reality of the universe we live in. I shall go into this in more detail; this point is too important for just a lick and a promise. But before beginning the argument itself, I want to briefly explain what sort of argument it is and what terrain it covers. The case is based on the laws ruling the world of molecules, the world out of which we have evolved. We will make an excursion into that world and search for natural time asymmetries—all the while, to be fair, keeping our intuitive notion of time at arm's length. With something like time, we cannot rely on intuition—time is too familiar, and the familiar is wont to make us blind or numb to the intrinsic reality.

Indeed, temporal asymmetry is so deeply ingrained in the ways one thinks about the world that it is very difficult to cut oneself loose from presuppositions. So we will keep intuition at a respectful distance and look for mathematically demonstrable time asymmetries in the world of molecules. Once we spot some, we will stake out their boundaries and set them over and against those of Evolution. And as we then reconnoiter the common ground, we will reap where we have sown: a time that is as pigheadedly directional as our inner time.

Let's begin with the molecular sphere. A search for asymmetries there won't take long. Two immediately stand out: the *cosmological arrow* and the *thermodynamics arrow*. The first was put on the map by astronomers toward the second half of the twentieth century, following the discovery that the galaxies are rushing apart with a speed approaching that of light (the distance between any pair of typical galaxies doubles every 10 billion years). The cosmological arrow points in the direction of that expansion—that is, away from the initial state, the Big Bang—and this is the direction it has taken ever since. Whether it will continue to do so forever is a matter of interesting uncertainty. This depends on the total mass in the universe, an unknown. If there is enough mass, the expansion should come to a halt (as the mass exerts its mutual gravitational pull) and be superseded by contraction ending in collapse, the so-called Big Crunch. According to data currently on hand, the universe contains only about 30 percent of the mass needed to halt expansion. As I write these lines, the results of observations of distant supernovas are coming in, suggesting that the universe's expansion may actually be speeding up, rather than slowing down. Thus the odds currently favor a picture of a perennially expanding universe.* But the jury is still out, and there is still much to be learned about the energy that is latent in empty space.

The second arrow, the thermodynamics one, points in the direction that the events in the molecular sphere normally unfold. Things there have not just an ordering in time, but also a direction: from high to low order. And this trend prevails in the entire coarse-grained universe— the *macro-domain* of physics. We are no strangers to this trend: cookies crumble, but don't reassemble; eggs scramble in the frying pan, but don't jump back into their shells; chinaware shatters, waves fizzle, trees fall to dust . . . and all the king's horses and all the king's men couldn't put them together again.

The plight of the Humpty Dumpties of this world is independent of the universe's expansion; in a contracting universe, the thermodynamics arrow would still point the way it does (it would do so even at the enormous spacetime warpage of black holes). The arrow represents a seemingly ineluctable trend and encompasses the entire sphere of molecules. The systems there constitute worlds by themselves, which can harbor immense numbers of units—a drop of seawater contains some 10^{19} molecules of sodium salt. Those molecules dance perennially to the thermals and tend to be scattered randomly about in what is called their equilibrium state. However, before they reach that higgledy-piggledy state, they show a penchant to deploy themselves in an orderly manner, forming aggregates—loose molecular associations, tightly bonded compounds, and so on—the sort of molecular states chemists use for their concoctions.

All those states are rather short lived. How much order a molecular system can embed depends on how much information there was to start

---

*"Expanding universe" means that the galaxies are rushing apart the way a dust cloud will rush apart when the dust particles are set in motion away from each other. The galaxies themselves are not expanding, nor is the solar system or space itself.

with and how much gets pumped in afresh. But information is something difficult to keep captive. It gives one the slip, as it were, and eventually will wriggle out of the strongest molecular bond; not even diamonds, which are all bonds, last forever. There is in truth no shackle, chain, or wall that could stop information from ultimately escaping—the thickest wall on Earth won't stop the transfer of heat or shield off the gravitational force. So in the long run, a molecular system left to its own devices becomes disordered.

That goes for complex systems, too, including us. The laboratory of bioevolution used the same sort of molecular aggregates for its concoctions as the chemists do. Our own organism holds immense amounts of information, and its systems are complex and tangled; but made of molecules as they are, their eventual decay to disorder is inevitable. We, and the other living beings, manage to stay that fate for a while by pumping in fresh information, and we do so ceaselessly and in enormous quantities to stay alive. Alas, time's arrow points its pitiless course day in, day out, and nobody can pump enough to keep the horseman with the scythe away forever.

How to Predict the Future from the Past

So much for the directionality of time and the distinction between past and future. As for the present, I intentionally left it out. Time flies continuously from being past to being future, and in such a continuum there is no room for a present. However, I do not wish to belittle what many hold so dear. I leave the present, and not without a little envy, to someone like the writer Luis Borges, who knew how to celebrate it so brilliantly. In quite a few of his tales the present holds center stage, and who could resist their spell when they bring a world to life, where the wish of things is to continue being what they are—the stone wishes to be stone, the tiger, tiger . . . and Borges, Borges.

Well, for us scientists, that "being" is but a "becoming," though we are perhaps the poorer for it.

But to get back to the directionality of time, its proximate cause lies in the world of molecules, the world out of which life has evolved.

There, fickle chance holds court, and systems are tossed together by the throw of the dice. Molecular systems—their component states, the molecular aggregates—are statistical in makeup and are therefore not too dependable. However, this doesn't mean that these states are totally unpredictable. They are not as predictable as those of the stars and planets in the celestial sphere, or those of the rise and fall of tides, or the fall of apples—not with the same degree of certainty. The movements of large objects are governed by Newton's laws, and his $F = ma$ predicts such events completely. That formula says that if the force F acting on a given mass m is known, its acceleration a is completely predictable. So, if the position and velocity of a thing can be measured at a given time (t), all we have to do is apply the rules of calculus to compute its position for a subsequent time, $t + dt$. Each of the locations then is completely determined by the ones before—they are specified for all times.

Such determinacy is something satisfying—it betokens regularity and order, which appeal to the human mind. Our mind abhors disorder and is perennially searching for regularities, even though such things may be scarce outside the inner I. That is an old yen, as old as human history. Folktales and traditions abound with personifications equating order with good and light: in Greek mythology the primordial world was one of disorder and darkness; the Bible speaks of formless darkness until "there was light!"; and the heroes in the Nibelungen Ring never seem to tire of battling the forces of darkness and of seeking the light. Well, when in years to come science will be lore, those heroes may still be at it, though their leitmotiv will have mellowed to something like "The light is in the regularities, in Nature's laws" (and be sung a merciful decibel lower).

Our brain somehow seems to be wired for such regularities—aeons of evolution have seen to that. All this development was given momentum by one basic need: *to know and understand what happens around us.* From there, it was but another step *to wanting to know what will happen before it actually happens,* and Newton's laws filled that desire more fully than anything before then. Indeed, his $F = ma$ is better than any crystal ball. It not only predicts the location of moving things, as we saw previously, but when coupled to another of his formulas, a law governing the force whereby two things at a distance attract each other, it can predict all sorts of events, like the paths of planets or comets, the wobble of the moon, the trajectory of a missile, the mass of the earth, the gaps in Saturn's rings, or the fall of a stone. When Newton presented the formulas to the Royal Society of London in 1686, he could say with justifiable pride: "I now demonstrate the frame of the System of the World."

It would eventually turn out that this frame bounded only a part of the world. But that's beside the point. What matters is that within those bounds, the formulas' predictions were infallible and still are—*within those bounds, the future is completely determined by the past.*

Now, as we descend to the molecular realm, some of that infallibility is gone. Here we encounter shades of determinacy, degrees of certainty, depending on the amount of information on hand. It is not possible to describe the individual motions of every single molecule, let alone predict them. The complexity is enormous. Say you try to describe the behavior of just a few molecules out of the zillions in a gas or a liquid. Those within that small group may follow straight paths for a short period, then bounce off each other in ways you may be able to predict from their previous paths. But just as you begin to discern a pattern, a molecule from another group zips along and crashes the party, breaking the pattern. And before you can blink an eye, along comes another molecule, and another, and another. . . . The complexities with so many moving pieces are overwhelming.

Nevertheless, if enough information *is* on hand, one can still pin down the states of a system with reasonable accuracy. Given some information about the significant variables in the molecular throw of dice, like position and speed, temperature and pressure, or chemical constitution and reaction rate, Newton's formula still works its magic. One has to bring in statistics to cope with the vast numbers of molecular pieces. But this is not too difficult—that sort of statistical mechanics in the gross is rather routine nowadays. Unavoidably, one loses some definition here—that's in the nature of the statistical beast. But one brings home the bacon: the probability of the system's states, or to be precise, the probability amplitude (which is simply the square of a measured quantity or the sum of such squares, when the event can occur in more than one way). And when that probability is high enough, it still can claim the dignity of being nature's law.

Thus, despite the turmoil and hazard of the die, we can still discern the regularities and predict events in the molecular world—the blurredness vanishes in the focus of the mathematics, as it were. There is a deficit of information in the molecular sphere, but that deficit is often small enough, so that *the future is determined by the past in terms of statistical probabilities.*

Why the Cookie Crumbles

This statistical future is precisely what the thermodynamics arrow chalks out. And given enough information, we can describe the various underlying statistical events. But the overall direction of those events, the arrow's direction, we see intuitively without mathematics, as we know the ultimate state: complete disorder (the equilibrium state wherein the molecules just ride the thermals and move randomly about). So in a bird's-eye view, the thermodynamics arrow of a system always points from the less probable state of order to the more probable state of disorder.

A loose analogy will make this clear. Consider a box containing a jigsaw puzzle whose many individual pieces have all been neatly put into place. If you shake the box, the pieces will become more and more disordered. Here and there, some of them may still hang together or fall into place, constituting parts of the original picture, but the more you shake, the more the pieces get jumbled up—it's the hand of time and chance.

There is no great secret to this hand. It is just the way things statistically pan out, because there are so many more ways for the jigsaw pieces to be disordered than ordered. Just weigh the collective states of order available in the box against those of disorder: there is only one state in which all the pieces fit together, but there are many more where they don't. So it is highly probable for such a closed system to become disordered, and the more pieces the system contains, the more probable this becomes. For molecular systems, wherein the pieces number zillions, that probability gets to be so overwhelming that it constitutes a dependable rule: the second law of thermodynamics.

We can see that rule at work everywhere around us: a crystal of sugar dissolves into a cup of tea, a lofty redwood tree molders into dust, the coiffure on which she worked for hours gets disheveled by one gust, sand castles crumble, rainbows become thin air, and the erstwhile *Titanic* is scoured with rust.*

Can Time Go Backward?

The reasons behind this run of things were discovered by the physicist Ludwig Boltzmann. He derived the Second Law from first principles, namely by tracking down the dynamics of molecular systems through a combination of statistics with Newtonian mechanics—one of the great theoretical achievements of all time. The equation subsuming that work of his (equation 1 in the appendix) has been fittingly inscribed on his tombstone. I had the opportunity to visit the tomb some years ago in Vienna; it lies not far from those of Beethoven, Brahms, and Schubert at the Central Cemetery. Boltzmann's paper appeared in 1877, but it has lost none of its luster and tells us more about time than a thousand clocks.

First and foremost, the paper shows that molecular systems evolve toward states with less order—and why they do: because there are so many more states with less order than with higher order. But between the lines, there is something that may, even a hundred years after Boltzmann, strike one as mind-bogglingly surreal: Evolution can, in principle, also go the other way round. Indeed, that point, namely full time reversibility, is implicit in Boltzmann's derivation, as it started from New-

---

*Not even the stars escape the trend to mortal dross. It's just a matter of time scale. Things up there don't bump into each other as often as they do in the molecular sphere, so the trend to disorder is hardly noticed on ordinary time scales. But it is noticeable on cosmic scales. Newton was quite aware of this when he formulated his laws of motion of stars and planets. "Motion is much more apt to be lost than got, and is always upon the decay," he said two centuries before Carnot's and Clausius's formulations of the Second Law.

ton's time-reversible laws. But the proof is in the pudding—a theorem in this case. And that came not long after Boltzmann's coup, when the mathematician Henri Poincaré incontrovertibly showed that any system obeying Newton's laws must, in due time, return to its original state (or to a state near it).

So the Queen's declarations on page 5, strange as they may seem, are really not so far out. They are certainly coherent and quite in line with physics theory. Indeed, they are in accord with the precept every physicist holds dear: that the universe is symmetrical. Symmetry is a modality of order that beckons the human mind. But with physicists this attraction goes well beyond the usual three-dimensional mirror and rotational symmetries. Their symmetries can include the time dimension or even dimensions we cannot see or feel—symmetries that are hidden in the mathematics, as it were. The contemporary unification theories, like the string theories that work with seven or more extra dimensions, are shining examples.

I could cite many instances of time symmetries—the unification enterprise attempting to bring all forces of nature under one roof has been booming for two centuries. But perhaps nothing illustrates better how deep the symmetry creed runs with physicists than what Einstein wrote when his lifelong friend, Michele Besso, died: "Now he has preceded me a little by parting from this strange world. This means nothing. To us believing physicists, the distinction between past, present and future has only the significance of a stubborn illusion."

Time and Our Perception of Reality

To sum up, time's arrow *is* reversible. The whole physics edifice affirms it, and a theorem backs it—and no one can argue with a theorem. And yet . . . and yet, the doubt gnaws. We never see the arrow reversing; we never see time running backward. Why?

The answer once again is provided by Poincaré's mathematics: the probability of a reversal is dismally small. One would have to wait an unimaginably long time for a reversal to happen—unimaginably long even by cosmological standards.

You can check that out yourself. The times for such a reversal to happen in the molecular world—or "Poincaré recurrence," as it is called—go up exponentially with the number of molecules in a system: 10^N seconds for a system of N molecules. Which even for a simple system, like a drop of saltwater, comes to $10^{10,000,000,000,000,000,000}$ seconds or about $10^{1,000,000,000,000}$ years. And this is what I meant by "unimaginably long." The entire age of the universe, the time from the Big Bang to today, is but 10^{17} seconds or about 10^{10} years.

A Poincaré recurrence thus lacks biological significance and is a matter of indifference to our pragmatic sensory brain. In all of bioevolution, time's arrow has pointed in one and the same direction; there has been but a steady progression of time in the molecular world of our universe—no regression, no cyclic repetition.

That monotony has always been the way of the world for us—our reality. It's a reality we proclaim in a thousand-and-one conscious and unconscious behavioral ways. We feel regret, remorse, and grief, and we project these to the past; we have expectations, aspirations, and hopes, and we project those to the future. We seek our opportunities in the future, not in the past; we know without giving it a moment's thought that we can affect the future but not the past. And if somebody shows us a movie of a lake in which a stone lifts itself off the water and is hustled by hundreds of amazingly collaborative water droplets through the air, we immediately know it's the movie, not time, that is running backward.

Our perception of reality is born of myriads of sensory experiences. That in principle things could be different, that well-grounded physics insists that time could reverse and the future become past, is of no consequence to our pragmatic sensory brain. To *it* moments, minutes, hours, years, aeons have fled into the past, never to be recovered.

CHAPTER TWO

Information Arrows

Retrieving the Past

Our discussion about time and its arrow included spans as long as all of cosmic evolution. Of arrows of such haul, we have but a smattering of tangible experience—the arrows in our everyday sensory world are but minuscule segments of them. If we may ever hope to bring those arrows within our grasp, we must go beyond our natural sensory horizon.

Not long ago such transcending would have landed us in metaphysics. But nowadays this is not only workable but bright with scientific promise. Advances in technology have placed the starry canopy at our beck and call. Indeed, what twinkles up there is a rich mine of time, and it doesn't take much quarrying to find arrows commensurate with biological and cosmic evolution. We see the nearest star as it was four years ago, the stars at the edge of our Milky Way as they were 100,000 years ago, and the farthest galaxies as they were 10 billion years ago.

To gaze at the stars, I admit, is not exactly a garden approach in biology. Though this shouldn't come as a surprise; the recent merger of physics and astronomy has produced a spectacular enlargement of our time horizon. We are now able to reach far back into Evolution, farther than Darwin could ever have dreamed, to the initial moments of the Big Bang. The Bang itself, to be sure, is cloaked in darkness; at that instant all energy and matter of our universe were squeezed into a single point, and there we reach the edge of scientific understanding. However, just a little further off there are no inherent limits to knowledge, and that is what physicists and astronomers are taking aim at. The first millionths of a second after the Bang are now within our reach.

That one is able to retrieve information from times so long past is largely owed to a good understanding of the principles governing nuclear reactions and to precise measurements of the afterglow of the heat of the Big Bang—the microwave background of the cosmos. The best measurements of that background were recently taken from a NASA satellite perched between Earth and the sun, offering a rather unobstructed view of the cosmos. With unprecedented precision (1 percent), this gave us a picture of the universe when it was a mere 380,000 years old.

The mechanisms of nuclear reactions and the underlying principles are now well enough understood to figure out what matter may have formed in the first few minutes of the hot universe, and experiments in large particle accelerators, simulating the primordial heat, supply data from which one can calculate what may have happened during the first thousandth of a second and onward. At the time of this writing the experiments had even gone as far as simulating conditions prevailing at 10^{-25} seconds after the Bang (10^{15} degrees Kelvin). In these conditions electrons, protons, neutrinos, and quarks are found to spring from pure energy—they literally materialize, offering us a perspective of nothing less than the creation of matter!

There is some irony in that. Darwin used to admonish those who would idly speculate about the origin of life that they might as well speculate about the origin of matter. Well, the tables have turned. We can now speak of the origin of matter with more confidence than of the origin of life. Indeed, so great is that confidence that physicists and astronomers have reached a consensus rarely seen anywhere else in science, and they unabashedly call the picture of the early universe described here the "standard model."

The certainty is not so high regarding later things—there are still quite a few lagoons in our knowledge of what befell all those 10^{15} tons of matter after they came into being. But we can sketch out the major evolutionary steps, at least for our neck of the woods.

So, we start our journey through time as close to the Bang as we presently can, from the first hundred thousandths of a second when protons and neutrons were born of quarks. This marks the most elementary organizational level we now conceive of. From there the structuring went on to other hierarchies: atomic nuclei, atoms, the first stars, galaxies, later-generation stars, planets, molecules, molecular aggregates . . . living beings. And if we put all of these organizations on a time chart, we can see ourselves in the broad context (see figure 2.1).

Now let's spread out this chart a little to see how things unfold. In the first hundred thousandths of a second, the most elementary nuclear organizations, the ones now deep inside the atoms, take shape. By three minutes, atom organization is off to a flying start, as hydrogen and helium nuclei form from protons and neutrons. Complete structures of atoms, organizations of the sort we now have around us, won't happen until several hundred thousand years later, when the temperature in the universe will drop to 10^3 degrees Kelvin. Then the first stable atoms will assemble from the nuclei at hand.

All throughout, the universe is expanding, but in regions with higher-than-average densities, all-embracing gravity begins to exert its pull. It has little effect on individual atoms, but at the large scales of the cosmic gravitational field, the atoms are pulled together. From those local condensations, stars and galaxies emerge (while the universe as a whole keeps expanding). And then things become rife for more complex atomic creations. The nuclear organizations of hydrogen and helium I mentioned previously, those that formed near the Bang, are evolutionary dead ends. More complex nuclei cannot form by simply adding neutrons or protons to helium nuclei, or by fusing them, because nuclei with five or eight particles are not stable. To get over that hump, higher densities are needed than those prevailing in first-generation stars, and such densities only occur toward the end of the lifetime of a star as it runs out of nuclear fuel and collapses onto itself. Then helium nuclei can collide with enough frequency to generate the next stable structure: carbon (C^{12}).

And now we are almost there: from that carbon, oxygen (O^{16}) can be made by adding a helium nucleus, and with carbon, oxygen, and helium, heavier elements can be built up under bombardment by neutrons, a bombardment supplied by a collapsing star.

18

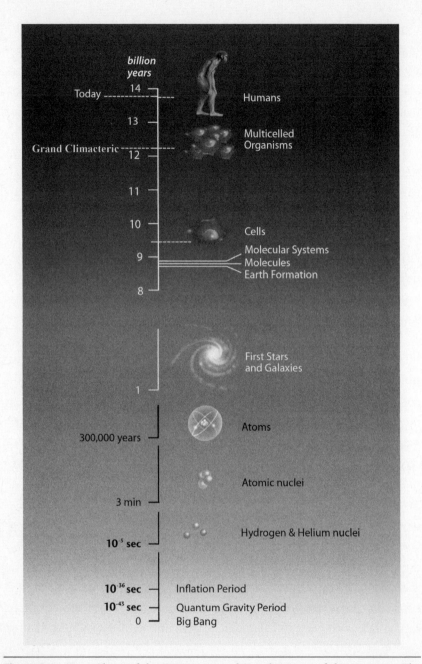

Figure 2.1. Time Chart of the Organizational Development of the Universe. The subchart on top contains the span from Earth's formation to "today," which is 13.7 billion years ± 1 percent) after the Big Bang, the time zero on the chart. The "Grand Climacteric" denotes the transition from single- to multicelled organisms; cells with nuclei and other organized organelles (eukaryotes) appear just below the mark.

If this seems long-winded . . . well, that's how it is. But where do *we* fit in? The carbon formation provides a hint, though the answer isn't obvious right away because atomic genesis is spread over such vast space and time. But things become clear as soon as we identify the organizational unit. This is of enormous dimensions: the galaxy—its stars (the offshoots of those "local" condensations I mentioned before) are all tied into one stupendous spiral whole.

In our case, the Milky Way is the organizational unit. And as we warm to the notion of such a spacious home, the answer comes: successive generations of stars here have transmuted the hydrogen born near the Bang into heavier atoms and seeded space with carbon, oxygen, calcium, iron, etc.—the building blocks of planets and life.

The Arrows of Time and Information

But there is something else that takes some warming to: the notion that there are billions of those organizational units scattered throughout the universe. By the latest count there are some 100 billion galaxies, so there may be many, many units out there where, from carbon 12, matter is forged and recycled. And each has its line of time. Those lines are not necessarily all exactly like the one in our own unit; the evolutionary steps leading from the basic building blocks to molecules, especially the complex ones, may well vary in different units. But one thing is sure under the Second Law: all the lines point in the same direction. And projected rearward, they all converge on one point: the moment when things were so close together that neither galaxies nor stars nor even atomic nuclei had separate existence (figure 2.2).

So we arrive at the origin, the initial condition that gave rise to the time asymmetry in our universe. The generally accepted notion is that at or near the origin, the universe was in a special state of order, which has been degrading ever since. The traditional way to deal with that singularity has been in terms of thermodynamics. But we will deal with it here in terms of information. That's the right and proper way in a book devoted to flow of information, though it may need some getting used to. But it really requires just a switch of train of thought, as information and

Figure 2.2. Information Arrows. Radiating out from the initial information state (*I*), the arrows bear for coordinates where organizations take place in the expanding universe. The small spheres on the arrows mark major way stations—galaxies, stars, planets, etc. Of the countless arrows in today's universe, only a handful are diagrammed here.

entropy are partner concepts. Thus, right from the start information, rather than energy, becomes the coin of the realm, offering the advantage of an outlook more directly in step with intuition, as we shall see.

Here then is a view of the initial conditions from the information angle: *The special state is the state of maximum information. It corresponds to the moment when the information of the universe was concentrated in a minuscule speck—all the information there was and ever would be. From this state emanates the information going to the units scattered throughout the universe and engendering their organizations. The respective streams of information mark the lines of time* (figure 2.2).

The starry canopy offers plenty of examples of such lines . . . and arrows. We can take our pick among the celestial arrows. Take, for example, that trusted sentinel in the northern sky, the North Star. Like all hot objects, it releases energy, and the direction of that energy flow is radially outward. This is what the word *radiation* stands for. But it also stands for the direction of the information flow. And information is what really matters to our sensory brain—it isn't energy per se, but information that makes us aware of the star. Energy is involved, of course, but only insofar as it embodies information and carries it down to us. The energy particles that perform this particular task are photons. They stream through space, hit our eyes, and transfer their package of information after some time. Thus the star's very existence implies an arrow pointing from the past. The arrow is a segment of a line of time.

So heaven's vault is crisscrossed with information arrows. The arrows hailing from out there are long—some have been on the fly for nearly 14 billion years. Those are the lines of information issuing from the primordial kernel, the initial state of information in the universe. Eventually that initial state led, in the course of the universe's expansion, to the condensation of matter locally and the formation of galaxies, as we have seen; as those vast structures evolved, more and more structures—stars, planets, moons, etc.—formed inside them.

From our perch in the universe, we ordinarily get to glimpse only segments of the arrows—local arrowlets, we might say. We therefore easily lose sight of the continuity. But as we wing ourselves high enough, we see that those arrowlets get handed down from system to system: from galaxy to stars to planets . . . to us.

The Products of Information Arrows: An Overview

It may seem surprising at first that there is organization at such a gigantic scale. If our imagination boggles at the number of cells in our brain, it will do so even more at the number of structures in outer space. There are some 100 billion galaxies, and each of these majestic spiral structures contains on the average some 100 billion star masses. Thus, some 10,000,000,000,000,000,000,000 stars are swarming out there, 10 billion times the number of cells in our brain. Each galaxy has an organization, and each star, in turn, its own.

All this bears witness to an inexhaustible multiformity in the cosmos, which is the product of information arrows just as the more familiar multiformity on our little planet is. And behind all of that, there is an informational continuity, a continuity of arrows stretching over an immense space and time (13.7 billion years).

Those are the arrows that engender the organizations in the universe. Each arrow has innumerable branches, whose prongs reach into every nook and cranny. Together, arrows, arrowlets, and prongs constitute a branchwork of staggering proportions—so staggering that one gratefully seizes on any expedient enabling one to simplify so complex a reality. Thus, making the most of their common directional sign, I have lumped the arrows together.

Life's Arrow

Well, except one. This one I couldn't bring myself to toss in with the lot. It bears for coordinates 0,0,0 . . . and directly down on us (see figure 2.3). Perhaps on Mount Olympus it may be no great shakes, but among us humans this arrow surely takes the limelight. Although it ought to be said that over a stretch of its flight, the arrow isn't as clear as we would wish; a significant portion of the information in the leg from the origin to the galactic unit is carried by the elementary particles of gravity, the *gravitons*, the physics of which is still in a state of querulous unease. But that is a cross we'll have to bear.

However, the last leg, the one from the sun to us, is straightforward. There the bulk of the information is carried by dependable elementary particles, the photons. These photons have wavelengths between 300 and 800 nanometers (from the ultraviolet to the infrared), and our planet gets liberally showered with them—26×10^{24} bits of information per second, an enormous information stream.

Figure 2.3. The Arrow Bearing for Coordinates 0,0,0. The initial information state (*I*) and the major way stations (*small spheres*) are represented as in figure 2.2; (*S*), our sun.

And this is the stream from which we suckle—we and all life on Earth.

High up on the biological totem pole as we are, we don't do the suckling ourselves, but enlist into service molecules that have a knack for using information from these elementary particles efficiently—a knack not many molecules on Earth possess. Prominent here are the chlorophylls of plants and the carotenes of bacteria (see figure 2.4). These are pigment molecules, and they are quite pretty—the chlorophylls are green, and the carotenes, yellow. Their bright color catches our eyes, but what really matters lies hidden inside them, in their electron clouds.

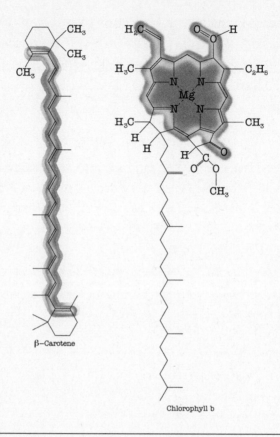

β–Carotene

Chlorophyll b

Figure 2.4. Windows of the Biomass—the Primary Photon Traps. The chemical structures of chlorophyll b and β-carotene; their photon traps— the regions with alternating single and double bonds—are shaded. These molecules are tuned to wavelengths between 400 and 500 nanometers— between the violet and blue/green. (The rest of the visible spectrum is covered by other molecular traps of this sort, which are not shown.)

There, around the carbon chains or carbon rings, a set of electrons is critically poised to absorb the photon energy quantum. The electrons lie in wait for that quantum, so to speak.

The pigment molecules are of modest size. They are not as large and fancy as their young cousins harvesting elementary-particle information in our brain, which we will meet further on. They are much older—some go back three billion years—and simpler. Chlorophyll has but a few dozen atoms, and β-carotene, even less. β-carotene consists of just a single carbon chain, with 11 electron pairs forming a photon trap. Nevertheless, both molecules are extremely efficient catchers of photon information—Evolution here hit early upon the perfect energy-and-information match.

The Strange Circles at Coordinates 0,0,0

Once the arrow is inside the biomass, it enjoys a long and eventful after-life. Its photon information first is stored in chemical form and then trickled down along chemical chains to where molecular organizations occur. And there is where it's at—where information begets order. Energy gets into the act here also, of course, but only insofar as it provides the driving force for the transfer of information. The begetting is the exequatur of information.

So, true to our colors, we'll skip blind-folded over the intricate details of those chemical chains, their local energy reservoirs and fluxes, and just concern ourselves with how the information flows. Indeed, all we need is a bird's-eye view of the flow, a broad perspective such as an observer from outer space might get. And there is something that is bound to catch his eye: the information ar-row, instead of going straight, as it does most everywhere else in the universe, goes in cir-cles here. Our observer will dutifully note that things are unusual at coordinates 0,0,0. He has no way of knowing what exactly is going on down there, but he is privy to the laws of physics governing the universe and will take it for granted that whatever it is, it can afford the steep thermodynamic price of going in circles.

We itch to tell him that the *it* is life, that information cycles in the macromolecular

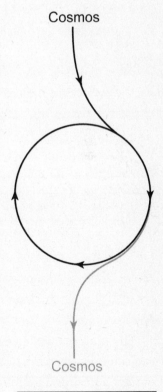

Figure 2.5. The Loops at Coordinates 0,0,0.

domain, but all the response we might get is an interrogation mark. So, unable to give a cogent definition and not a little embarrassed, we send him a few samples of the molecules that are responsible for the circular information flow. And to reassure him that there is no hocus pocus, we append a brief, explanatory message about how the thermodynamic price gets paid.

Molecular Demons

We'll get to the message shortly. First, a word about those macromolecules. These are modular, three-dimensional structures ranging in size from a few hundred to thousands of modules. Their atoms are so organized that the modules can quickly redeploy in response to electromagnetic forces, causing the macromolecules to swiftly change their shape. And herein lies their power, for they can bring to bear weak electromagnetic forces (van der Waals interactions) on other molecules, constraining them to take on determinate spatial orientations.

Now, to do the constraining securely and with a modicum of stability, the surfaces of the interacting molecules must be in close proximity to each other. In the case of three-dimensional molecular structures, this means that the partner molecules must fit like hand in glove. Thus the macromolecule, in effect, recognizes its partner—it selects it out of a myriad of other molecules. This is what I meant by "power." I meant information power—power of cognition.

It may seem odd to hear the word *cognition* in a molecular context. The term is commonly used for higher-order psychological phenomena, like the recognition of a face or an object. But the essence is the same: a selection among several choices. And bear with me a moment; we shall see that in either case the choices follow the same physics rules and are bounded by the same limits.

The aforementioned macromolecules and their retinue of fitting molecules form the backbone of the information lines inside living beings. Such lines are inherently noisy; they are largely in liquid state, and information transmission here relies on the molecules riding the thermals.* So things are not as simple as in telephone lines. There all one has to do is impress a gradient (a voltage), and the information carriers (the electrons) will run down the length of the wire. In the liquid-state biological lines, the information is carried by bulky molecules and transferred from one molecule to another upon random encounter. In such a haphazard communication, a gradient alone won't do. The pertinent molecules must somehow recognize each other, and that has its price. There is no free lunch in cognition—not even in photosynthesis. Cognition represents an information gain, and the stern laws of physics demand that information be brought into the line from outside to balance the thermodynamic accounts. And this is precisely what some of these large molecules do: *they draw the requisite outside information onto themselves and funnel it into the biological information lines.*

Now that is a talent nowhere to be found outside the orb of life. All molecules, including the inorganic ones, can carry information—the information intrinsic in their structure. But few even among the macromolecules can summon the *extrinsic* information needed for the thermodynamic quid pro quo of cognition. The simplest macromolecules with that talent have nucleotide modules. They are small RNAs with only a few hundred modules, which can switch from one- to three-dimensional information form. Those molecules came early onto the biological scene, and they probably were the first to be up to the cognition task (they can still be found at the lower zoological rungs and in our own ribosomes). Today, molecules with such talent are a dime a dozen in biological systems, and the predominant ones are made of amino-acid modules, usually thousands of them, complexly deployed in three dimensions—the proteins.

One is not used to thinking of molecules, certainly not the molecules physicists ordinarily deal with, as having smarts. One physicist, though, would have felt quite at ease with such oddities: James Maxwell, the

---

*I deal in detail with these lines and the mechanisms of their molecular information transfer in *The Touchstone of Life* (New York: Oxford University Press, 1999; London: Penguin Books, 2000).

towering figure of the nineteenth century. Maxwell gave us the theory of electromagnetism, and he also was the first to see the regularities amid the tumult in the molecular world—the regularities of statistical averages. In his lifetime those macromolecules were not known, but had they been, he would not have been in doubt about what to call them. In 1871 he published a famous gedankenexperiment considering a hypothetical sentient being—a "demon" in his words—who could see molecules and follow them along their whimsical course. That demon would engage the thoughts of scientists for a century. An artful dodger of the Second Law, the demon was able to distinguish individual molecules and pick them out from a mixture, creating order from chaos.

Well, the macromolecules I mentioned have exactly the talent of Maxwell's demons: *they can pick a molecule out of a universe.* But there is this little difference: they do so entirely within the law.

From now on I will call those macromolecules demons for short. They have that talent for selection built right into their three-dimensional structure. Take a protein demon, for instance. He has a cavity that fits the contours of the molecule of his choice, and he holds fast to it through electrostatic interaction (see figure 2.6, item b). The electromagnetic forces involved here are weak and operate only over very short distances, 0.3–0.4 nanometer. So only a molecule that closely matches the cavity's shape will stay in place—one molecule out of a myriad.

Wresting Information from Entropy: The Quintessence of Cognition

But there is more to the demon's act than sheer force. The electromagnetic forces in the demon's cavity entrap the molecule of choice, but that's not enough for recognition. Not even the best mousetrap could by any stretch of imagination be called a cognitive device. Cognition is something infinitely more subtle. We try to come to grips with it little by little in this book, as we move up the information trail in the brain. But here let me say that, at the very minimum, cognition entails an information gain, a net profit—and our protein demon reaps it through a

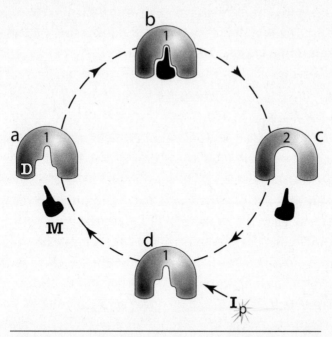

Figure 2.6. The Cognitive Cycle of a Protein Demon. The demon (*D*) and his molecular mate (*M*) go through a cycle (a→b→c→d→a) in which the electromagnetic interaction with the mate triggers the demon to switch from gestalt *1* to *2*; information extracted from organic phosphate (*Ip*) allows the demon to switch back to gestalt *1*.

skillful thermodynamic transaction. Much of that transaction is hidden to our eyes, and it takes the information lens to spot it.

To begin with, let's see how the entropies flow as the demon reaps his profit. His possible information gain is strictly regulated. It is governed by the first and second laws of thermodynamics, depending on the total entropies produced, and cannot exceed a certain value. But because our sun is such a huge extrinsic-information reservoir, the gain, for all practical purposes, is limited only by the entropies produced inside and outside the information system. Or put differently, the gain in information ($\Delta I\mathbf{m}$) could, in principle, be as large as we wish, so long as it balances out the sum of the associated increases of entropy; to wit, the entropy in the molecular system of which the protein demon is a part (ΔSm) and that in the environment outside the system (ΔSe). In mathematical shorthand:

$$\Delta Im = \Delta Se + \Delta Sm$$

This is the message we would send to the observer in space, along with the molecular samples, to set him at ease that those molecules strike that entropy balance. That equation provides the quintessence of cognition (hereafter *cognition equation*). And for good measure, we would append a few milligrams of ATP (adenosine triphosphate), which most biomolecular demons on Earth use as a surrogate of the sun.

The rest our observer will find out himself by watching the molecular interplay. Though it won't be easy—the demon juggles fast. He twists and wriggles as his amino-acid modules swivel, causing the molecular configuration to change in a fraction of a second. But that's mostly sleight of hand. For all his writhing, he just switches between two gestalts: in one, his cavity is open; in the other, it is not. He methodically goes through a cycle. In gestalt *1*, he is ready to catch the object of his desire. But no sooner does he score, than he shifts to gestalt *2* and bids his mate good-bye. Then he reverts to gestalt *1*, ready for more of the same (figure 2.6).

It is then and there that the demon strikes the mandated entropy balance. He pays the entropy price with the bits he palms from the ATP— and he does so as he reverts to gestalt *1*. The two gestalts are quasi-stable states of molecular structure, so it is easy to go from one to the other, somewhat like a see-saw. It just takes a little extra to tip the balance (see figure 2.7). A tiny force, like the one produced by the electromagnetic interaction with the mate, will do the trick in the *1→2* direction. But in the opposite direction, for the resetting of his structure, the demon needs to be primed anew. And here is where the ATP comes in.

That phosphate is a chemical storage form of energy from the sun that organisms use widely for their chemical operations. The solar energy is concentrated in the phosphate tetrahedral bonds, and the demon breaks those bonds with water, setting their energy free. But it is really

Figure 2.7. The Little Extra That Tips the Balance.

not the energy he hankers after (although you might think so by reading biochemistry books). To be sure, there is a good deal of energy stored in those phosphates, and as the tetrahedra break apart, all this energy goes off in one blast (7.3 kilocalories per mole). But much of that is just flim-flam. Much of the energy goes down the drain; it merely heats the environment. Indeed, all of it would do so if the demon weren't there (and the tetrahedral bonds were broken by water alone).

So the demon's secret stands revealed: he palms ATP and nabs information from it that otherwise would go to waste. His trick—and it is the most dazzling of magician's acts—is to wrest information from entropy!

Who Shoulders Life's Arrow?

That, then, is the feat a molecular demon brings off in each information round as he changes gestalt, and why he can claim the dignity of a cognitive entity. A protein demon typically goes through hundreds or thousands of rounds—or let's call them by their rightful name, cognition rounds. At the end of each round, he makes his bow to the thermodynamics law and antes up the mandated information. He draws that information in from outside—he and countless other demons in an organism. In the aggregate, those demons bring in a gargantuan amount, a sum by far exceeding the amount coming from the DNA. This is all too easily overlooked, as one naturally tends to gravitate in biology to the DNA, the spectacular repository of organismic information. However, that is but information in two-dimensional storage form—it's something that sits on the throne, so to speak. To get it to do work, you need an altogether different kettle of fish: something that can move and act out the information in the world of three dimensions in which life evolved. This is a tall order to fill because, apart from the dimensional problem, the "acting out" entails thermodynamically costly cognition. But the protein (and ancient RNA) demons have what it takes: the internal information, the three-dimensional agility, and the wherewithal to shoulder the cost.

They are the ones who shoulder the arrow at coordinates 0,0,0. And we now also can justify the name we used for it before, life's arrow. That arrow has been engendering forms of ever-increasing complexity and order over the aeons—and I use "engendering" in the sense of immanent cause, as I did before.

But my purpose here is not to unfurl the flag. It is to show the continuity of the information arrow and the evolutionary continuity between the cosmic and biological. It is difficult to say where the one ends and the other begins. That is the nature of the evolutionary beast. But for the sake of argument, let's assign the beginning of the biological part to the emergence of the first three-dimensional macromolecular (RNA) demons. Others may prefer to begin their count of biological time later, from the emergence of the small photon-trapping pigment molecules. On such long scales, it hardly matters from which of these points we start. Either way we get about four billion years. And onward, if all goes well, another four billion years are in the offing. By then our sun will have burned out, and when that photon source dries up, there is no help for it—life's arrow will cease on this planet.

CHAPTER THREE

The Second Coming

We turn now to another arrow, a late arrival on Earth. This one came from the same bow as the first, but it had been waiting in the wings for over a billion years, biding its time for a fitting target, as it were (see figure 3.1). Meanwhile things had changed down there, calling for a faster transmission of information in the biomass. For aeons that transmission had been carried out by atoms and molecules moving randomly in water. But with the advent of multicelled beings, the need for a faster mode of transmission arose, and Evolution met it with a change in strategy: she would use the cell membrane as a medium for diffusing information, rather than only the cell water. Thus molecular information now was constrained to move inside a thin film, or even in single-atom file, instead of doing so freely in the bulk of water inside and outside the cells.

Such a modification of strategy doesn't happen overnight. It is no easy thing to change from free and easy wanderlust to goose step. We can roughly chalk out the turning point at somewhere between 1 billion and 700 million years ago on the cosmic time chart (figure 2.1), when certain cells grew sprouts capable of conducting electrical signals. These were digital signals—brief (millisecond) pulses of invariant voltage—that were transmitted along the cell membrane at high speeds (on the order of 1 meter per second, and eventually up to 100 meters per second), several orders of magnitude higher than the speed of signals in cell water. And those signals were strong (a tenth of a volt), well above the electrical background noise.

Figure 3.1. The Second Arrow.

Given this uptrend, organisms would eventually become criss-crossed with new and faster communication lines. Indeed, a whole new web would come into being—the neuron web—which used the digital electric signals for both communication and computation.

The Demons for Fast Information Transmission

As before, the leading actors in the new regime were proteins. Though these were special: their subunits were symmetrically deployed about a central axis, leaving a narrow water channel in the middle (see figure 3.2). The ones in use today have four or five subunits. They are osten-

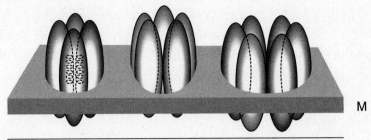

M

Figure 3.2. Channels in Cell Membranes. They are made of four, five, or six protein subunits; the α-helices from two neighboring subunits are roughed in in the four-subunit channel. The ion-selective channels belong to the four-subunit type. The least selective is the six-subunit type, which occurs at cell junctions; this channel has the widest bore and admits cellular metabolites in addition to inorganic ions. (After Unwin 1989.)

sibly designed to funnel small, inorganic ions through the membrane in which they are embedded—chloride, potassium, sodium, calcium, and the like. Such ions carry a negative or positive charge. So, when the channels momentarily open up, a pulse of electric current flows through the membrane.

But there is more to this than can be gleaned from mere externals. The most intriguing things happen in the tortuous, narrow spaces between the protein helices lining the channel. There, certain ion passengers get picked out of the crowd by the protein, while the protein itself changes gestalt. This brings to mind the act of cognition of the molecular demons we saw before (figure 2.6). Indeed, here too, the protein methodically changes gestalt, switching from channel-closed gestalt to channel-open gestalt to channel-closed gestalt, and with each cycle the membrane system to which the protein belongs reaps an information profit. And if we focus our information loupe on the atoms lining the narrow parts of the channel, we spot the cloven hoof: an electromagnetic field implementing the cognition of the ions filing through.

Several types of such demons are found in cell membranes. They differ in structure and cull out different ions—what is chaff for some is wheat for others. Those biases are the results of a long, long series of experiments in the laboratory of Evolution, in which α-helices and electrically charged (polar) amino-acid units were reshuffled and their spatial disposition fine-tuned. It was a relentless search, a single-minded quest for channel structures that would produce signals strong enough to stand out from the background noise.

At first sight, that may seem a straightforward task. But it's only the selection criterion, the signal-to-noise ratio, that is straightforward. Actually, things are perversely entangled, because in the membrane system as a whole, that ratio depends on two opposing factors: the rate at which the ions flow through the channel (flux) and how well different ions get sieved out (selectivity). On the one hand, to get a strong signal the flux has to be large—which calls for a wide channel. On the other, a wide channel has little selectivity—and without selection, there is no signal, as the flow then merely shunts the electrochemical gradients across the membrane.

It's one of those damned-if-you-do-and-damned-if-you-don't binds. But Lady Evolution rose to the occasion—well, she always does in this book. She made the channel narrow, but only over a very short stretch, thus achieving ion selectivity with minimal impediment to flow—a Solomonic compromise.

Precisely how many trials she needed to steer that middle course, we may never know. The failures in her laboratory—and given the magnitude of the problem, there probably were many—have been wiped off the record. That's the nature of the beast. But the winners are here. They are all over the biological map, and their performances speak volumes. Their fluxes range between 1 million and 10 million ions per second per channel—fluxes like that virtually guarantee a strong electrical signal. And the selectivities, as measured in three types of protein channels, are 10:1 for sodium over potassium or calcium ions, 1,000:1 for calcium over potassium or sodium, and 10,000:1 for potassium over sodium—and in this sense one speaks of *sodium channels, calcium channels*, or *potassium channels*.

Let's take a closer look at that short stretch where the channel narrows. That stretch holds the key to the secret of how a demon tells the ions apart. These are quite small—much smaller than the molecules interacting with the garden-variety demons we dealt with before. It's out of the question therefore that he could use for his cognition long-range electromagnetic forces (van der Waals interactions), as the demon in figure 2.6 does. Those forces are too weak to hold onto a tiny potassium or sodium ion. Besides, those ions differ in size by just a fraction of an angstrom (1 angstrom = 0.1 nanometer). Yet a good channel demon manages to tell them apart with few errors. Just to cite one example, the demon in the membrane of a lowly bacterium does so with an error rate of 0.0001 and achieves fluxes of 10 million potassium ions to boot.

How does this demon bring off that astonishing coup? His secret came to light only a few years ago, when the biophysicist Roderick Mac-Kinnon and his colleagues succeeded, by X-ray diffraction, in getting snapshots of such a potassium channel at near atomic resolution. The snapshots showed that the ions move in single file through the narrows of the channel. And "narrows" is the right word. They leave room for just the naked potassium ion, the ion without its usual shell of water. Moreover, the narrows are studded with oxygen atoms (which are part of the protein backbone of the channel), and these can bring to bear strong, short-range (covalent) forces on the ion (see figure 3.3).

Thus we begin to see the trick the cunning demon has up his sleeve: he makes the potassium ions run the gauntlet in the oxygen-spiked narrows of his channel and offers them an oxygen atom from his own carbon-channel structure (a carbonyl oxygen) in exchange for a water oxygen atom. And he puts no effort into it. He doesn't need to. His channel is tailor-made for that. The radius of the narrows—1.33 angstrom—is exactly right to pull the water from the potassium ion, and the ion fits so well that the energy costs and gains in the exchange are balanced. (On the other hand, an ion as slim as sodium hasn't much of a chance to get in on that give and take—the narrows are too roomy for the ion to get close enough.) Moreover, the mutual repulsion between potassium ions balances out their binding to the channel; when two potassium ions get within 3 angstroms of each other, as they do going through the narrows in single file, the repulsive force between them is just right to overcome their binding to the channel. And so a potassium ion entering the narrows in the wake of another will knock the first one off its binding site, and so on.

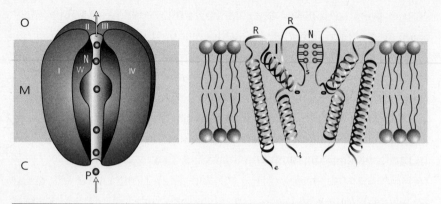

Figure 3.3. A Potassium Channel. *Left:* A cutaway view of the protein and its four subunits. (*W*) water cavity and (*N*) narrows of the channel, with a file of potassium ions in transit; the arrows indicate the direction of transit. (*M*) cell membrane, (*C*) cytoplasm, (*O*) cell exterior. *Right:* A schematic of the α-helices and their backbone ribbons (*R*) in two contiguous subunits. (*S*), carbonyl oxygen of the narrows. The inner helices (*i*) face the channel and the outer ones (*e*) face the lipid bilayer. The linking helices (*l*) act as dipoles, stabilizing the potassium ion inside the water cavity; (-) is their negatively charged carbonyl end. (Drawn after 4-Å Fourier maps of *Streptomyces* channel crystals and other X-ray data by Doyle et al. 1998 and Gulbis et al. 2000.)

To give the movement of these ions a direction, the demon takes advantage of the prevailing electrochemical potential across the cell membrane (cell interior negative). That potential makes a good driving force and evolutionarily comes as a windfall for the demon. It was there long before he came on the biological scene—it dates back to single-cell times. The demon uses it judiciously. The electrical component of the potentials gets used up mainly in the narrows of his channel, the final stretch of the ions' journey through the cell membrane. Here is where their movement meets with the highest resistance,* and much of the energy dissipates as heat. But even those energy losses are

---

*Elsewhere, the channel is relatively wide; it is lined (including the cavity) mainly with hydrophobic amino-acid residues, a rather inert surface. The linking helices (*l* in figure 3.3) are polar—effectively, dipoles. Their negatively charged ends point toward the channel's cavity, stabilizing the potassium ion in the water there.

minimized, as the potassium ions push away from one another and a good part of the resistance gets nulled by that mutual repulsion.

The Sensory Demons

The design principle of this demon probably applies to quite a few other ion-selective membrane channels, including those that interest us most in this book, the potassium channels of the neurons of our brain. There may be a little more elbow room inside those neuronal channels, so that a few sodium ions slink through—the error rate is higher, 0.001. But a little cognitive sloppiness is not necessarily a bad thing in molecular demons; it speeds up the interactions with their customers. In any event, it is tolerable in this case because the safety margins are sufficiently high. The fact is, these demons still produce electrical signals loud enough to be heard from toe to brain.

But if their lower information power is a shortcoming, it is more than made up for by their workmates in the sense organs. In those organs, the channel demons tend to associate with another class of membrane demons, the *sensory demons*, and these are as discriminating as any biomolecular demons we have seen. They will pick a single molecule out of a sea of others—some of them, the most spectacular ones, will even pick out a single photon.

The sensory demons constitute a large and variegated class. They are strategically located at the brain's sensing outposts—the sensory systems in our eyes, ears, nose, tongue, skin, and so on. These systems are but extensions of the enormous conglomerate of nerve cells we call the brain, and their sensory demons bring in much of the information that the brain uses in its daily operations. Those demons are the brain's antennas to the outside world. Previously in this chapter I mentioned the suitable targets the Second Information Arrow was waiting for at coordinates 0,0,0. Well, here they are. The sensory demons are the targets of that arrow—its factotums on Earth.

In a cosmic perspective these demons are thus the functional equivalents of the chlorophylls and carotenes of yore, which originally let the Information Arrow into the biomass. To us humans, though, they seem so much more, as they allow us to become aware of the universe. They feed our brain with information about the world around us, its endless hues, its shades of color, its melody of sound, its velvets of touch, its flavors and smells.

A Generalized Sensory Scheme

I expand on that in the following chapters, as we work our way from the sensory periphery up to the centers of the brain. Here we focus on what happens at the brain's sensory outposts, on the local action of the sensory demons and the interplay with their channel workmates.

Perhaps I should say partners, for theirs is a partnership in the best sense of the word; the sensory demons attend to the business of gathering information from the world outside, and the channel demons, to the business of digitizing and conveying that information in digital electrical form to the centers of the brain. That form is the same in all sensory outposts of our brain—all our senses, regardless of what kind of sensation they produce.

The two partners maintain a degree of independence, which has important consequences in the way our brain represents the world outside. More on that later. But if we consider their actions overall, we find that they are well-coordinated. So our immediate question is, how does the coordination come about? And since coordination implies information transfer, we ask straight out, how do the two demons communicate?

The answer depends on how far apart they are. In our simplest mechano-receptors, the sensory demon generates an electrical signal himself (generator current), although not a digital one. The digitizing is the purview of his channel partner. But as the two demons sit side by side here in the neuron membrane, a communication by a direct electrical interaction will do. This is not so in most other receptor systems. In many the two demons are not within electromagnetic earshot of each other—in some not even in the same cell membrane. Our visual system provides perhaps the best-known example; the sensory demon is located in an intracellular membrane and the channel demon in a cell surface membrane, a gaping gulf of cell water intervening between. Electrical communication thus is not an option. The sensory demon here bridges the gulf by sending messenger molecules to his partner, molecules that can diffuse in cell water.

That by itself wouldn't be surprising—communication by diffusible messenger molecules had been in vogue since life began. But what is surprising is that the messenger molecules between demon partners in widely disparate receptor systems are strikingly similar, if not the same. The photon-sensing demons in our eyes use messenger molecules quite similar to the ones the pituitary hormone–sensing demons in the ovaries and testis use!

Such sweeping likeness warms the cockles of the reductionist's heart. It speaks of a common evolutionary design theme—and this one seems to go way back to when brains were not yet around, which promises operational simplicity. Indeed, under the information loupe, the primary receptor operations in senses as diverse as our visual system and olfactory system or the primary receptor operations in the basal parts of our brain all reduce to a simple scheme: two protein demons tethered to their respective membrane sites and a set of diffusible molecules (I_{m1}, I_{m2}, ... I_{mn}) shuttling in between. These molecules constitute a relay chain wherein one member transfers information to another by direct (largely van der Waals) interaction. In shorthand,

$$\text{Sensory Demon} \rightarrow I_{m1}, \rightarrow I_{m2}, \rightarrow \ldots I_{mn} \rightarrow \text{Channel Demon*}$$

This scheme delineates the primary information flow—the mainstream from the vantage point of the brain. There are other flows in addition, accessory inflows. There have to be; the thermodynamic books need to be balanced out. That's the law. The extrinsic information mandated by the cognition equation is fed into the various links of the chain. And that infusion is usually in the form of high-energy phosphate, which the I_ms themselves often carry to the links piggyback. In higher

---

*The various I_ms denote molecules diffusing in cell water or in cell membrane. Many are derivatives of guanine. The reader interested in the biochemical details will find those in physiology and molecular biology books.

sense organs, such accessory flows are many and, if we were to heed them all, our diagrams would become crowded. However, keeping our loupe duly focused, the essence shows through the legerdemain: the flow *between* the demons. And the fact that all the I_ms are in series here gives us leave to do some more streamlining and to lump them into one term, I_M. And so we arrive at a general scheme:

$$\text{Sensory Demon} \rightarrow I_M \rightarrow \text{Channel Demon}$$

The Demon Tandem

Now, shifting our line of inquiry from demon communication to demon behavior, the picture becomes more familiar. The two demons change gestalt as they go through their rounds of cognition. As diagrammed in figure 3.4, the sensory demon undergoes a change in molecular conformation in response to an adequate stimulus—an elementary particle (a photon) in the case of a visual sensory cell or a molecule in the case of an olfactory cell or a hormone-sensing cell. Starting at the demon's receiver region, that conformational change propagates rapidly through his structure. And as a result, the demon releases the first information carrier in the chain (I_{m1}). This molecule, then, diffuses to the next in line, and so on. The last eventually interacts with the channel demon, who, responding with a conformational change of his own, opens up and lets the ions through.

The example in figure 3.4 represents a relay chain with three members. That number varies from sensory system to sensory system, but in all cases the systems use the most economical means for getting the molecular information from one place to another: the thermals. All the molecules in the chain move *randomly*—in the cell water or the cell membrane, as the case may be—and pass information to one another upon *random* encounter. But all the same, they do the job; there are enough copies of them around to statistically ensure informational continuity among the demons.

If we pause to reflect a moment on the purport of that mode of information transmission for Evolution's neuronal enterprise, we realize that,

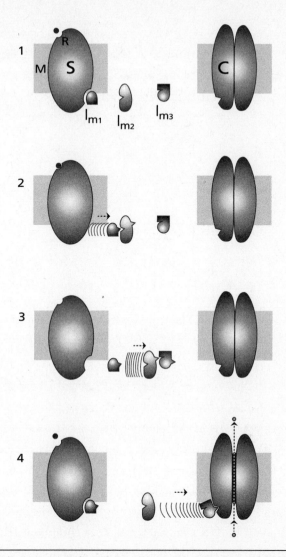

Figure 3.4. A Sensory Relay Chain. (*S*) sensory demon; (*R*) his receiver region (the black dot represents the adequate stimulus); (*C*) channel demon; (*M*) membrane; (cell exterior at top of membrane); (I_{m1}, I_{m2}, I_{m3}) members of the relay chain. The diagrams depict four stages of information transfer (*1, 2, 3, 4*). The participants in the chain successively interact at complementary surfaces, each inducing a conformational change in the next in line. In *2* the sensory demon has changed conformation in response to the stimulus (*black dot*), setting the chain in motion. In *4*, the channel demon changes conformation, opening up, while the sensory demon has completed his conformational cycle and is rebinding I_{m1} to go into standby (*1*). The horizontal arrows indicate the direction of the information flow in the chain. The vertical arrow through the channel indicates the flow of ions (*open circles*) in the case of a potassium channel.

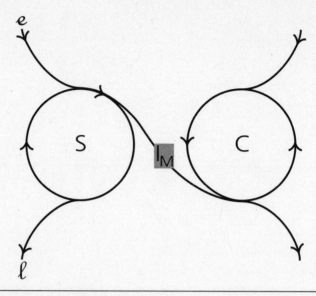

Figure 3.5. The Sensory Demon Tandem. Cognition cycles of the sensory (S) and channel demons (C) and their I_M linkage. The arrows indicate the direction of the information flows in the tandem. (e) inflow of extrinsic information—the information carried by the adequate stimulus plus that supplied by ATP or other energy sources, to balance the entropies. (l) outflow—the combined information losses.

apart from economy, there was another benefit to her choice of the random molecular mode: time asymmetry. Such a mode is inherently time-asymmetrical (see chapter 2), which would not have been the case had she chosen a purely quantum-transmission mode. Indeed, each link in the molecular chain here is asymmetrical, so I_M points a strong arrow, linking the cognitive cycles of the two demons in tandem (see figure 3.5).

It is this that makes a physiological system out of what otherwise would be just two freewheeling agents. While each demon here does his thing and, as if by rote, does so over and over, I_M turns them into a harmonious whole with a time direction.

How the Demon Tandems Censor Incoming Information

Those tandems occupy a strategic, even privileged, position in our brain. Located at the brain's sensory outposts, they sit in judgment on the information coming in from the world outside and decide what of it is worth our attention and awareness.

One might think that such an important decision would be left entirely to the higher centers of the brain. But it is not. Much of the information not worthy of our awareness doesn't even get that far. It is censored out by the demon tandems—stopped in its tracks at the very outposts of the brain. It's a censorship directed at the humdrum, at information about the world's status quo. The tandems simply won't put up with that kind of monotony and will uncouple, unceremoniously cutting off the information flow to the centers of the brain.

Figure 3.6 illustrates this point. It shows an example in which a tandem is presented with an invariant environmental stimulus, one that was rapidly applied and then held constant. Typically, the channel demon

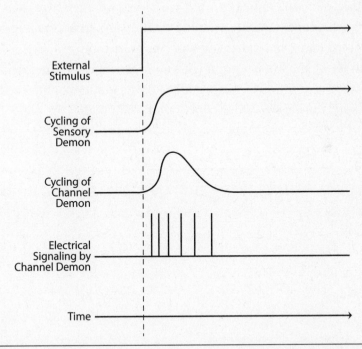

Figure 3.6. Automatic Uncoupling of a Sensory Tandem. Rate of cycling of the sensory and channel demon in response to a continuously sustained stimulus (time zero: onset of stimulus). The ensuing digital electrical signals are plotted on the same timescale; they are pulses of millisecond duration that are compressed to lines on this timescale.

starts his cognition cycle here as soon as he receives his partner's I_M message—and systematically goes through a number of opening and closing rounds. But then he stops in his own good time—he stays put in the midst of a cycle, no longer producing digital electrical signals—and this despite the fact that his sensory partner keeps cycling on.

The tandem uncoupling is at the root of what physiologists call *sensory receptor adaptation*. It's an ancient mechanism antedating brains by a billion years. It goes back to the beginning of the multicell era, when nervous systems consisted of just a handful of neurons connected to gland or muscle cells. Such systems allowed organisms to reflexively respond to fast changes in the environment—to jump when the ground got hot, to run at a rustle in the bush, or to pounce on what writhed or fluttered. They were systems designed to respond to environmental transients, not to the status quo.

In time more and more neurons were added to those systems, forming elaborate neuronal circuits. And one circuit was heaped upon another and twisted into a fantastic trellis with ever more complex functions. But even today, in the human brain with more than a trillion neurons, the primeval system shines through, and the primary interest of the sensory brain is still the sensing of transients, not the status quo.

CHAPTER FOUR

The Sensors

Now that we have become acquainted with the lead characters of the sensory tableau and the way they admit, and to some extent control, the Second Arrow, let's put them in a cellular context. They are located in specialized nerve cells, the *sensory cells*—in the cell bodies or the tips of extensions (dendrites) strategically situated to detect changes in the world outside or inside the organism. All characters here are protein demons, and all are ensconced in membrane—a rather greasy, oily setting. But that's what it takes to accommodate them and keep them happy. The bulk of their bodies is water repellent, and the membrane lipid holds them in place, yet offers them enough freedom to move. Besides, it provides good electrical insulation—a necessary condition for electrical signal transmission.

The Sensory Transducer Unit

Thus more than a setting, the membrane lipid is an integral part of the sensory process. It is all very well and good to treat them separately for analytical purposes, as we have done so far, but when it comes to physiology, sensory demon, channel demon, and membrane form a functional whole. They constitute a unit that converts environmental information into electrical signals, or, if you wish, a unit that converts environmental energy into electrical energy. Either way, it's a *transducer*, so we will now examine the sensory process in this light.

Even a priori, the transducer concept furthers our purpose. It offers a rationale for classifying the sense organs into three categories: *mechano-electric*, *chemo-electric*, and *photo-electric* transducers. This categorization may at first seem strange because it is so unlike the one we daily use or see in textbooks. But as we consider the full expanse of sense organs in the biosphere, we realize with a painful twinge that our habitual categorization—touch, smell, taste, hearing, seeing—only goes so far. Where, in those anthropocentric pigeonholes, would we put the rattlesnakes' sensing of infrared waves radiated by prey or the sharks' exquisite sensing of electrical fields?

The transducer pigeonholes, on the other hand, are both comprehensive and biologically meaningful—and, if you have a yen for demon taxonomy, you get that into the bargain. Take, for instance, the photon-sensitive demons. There are many types of these speckled over the zoological map, each covering a range of photon wavelengths. Three alone are in our eyes, transducing the visible wavelengths; others, in insect eyes, transduce the ultraviolet; yet others, in snake pit organs, transduce the infrared. Those infrared sense organs are unlike any of our own, as are the sense organs of skates, rays, and sharks, whose demons transduce the electric fields in the sea.

Or take the mechano-electric-transducer pigeonhole. That one houses an even larger demon crowd—it's older, going back to single-cell times, well before the advent of the channel demons. But we still can find some of the back numbers in bacteria and protozoa, practicing the sober jog trot of old times: transducing hydrostatic or osmotic pressure. The modern demon generation is more sophisticated. It went into partnership with the channel demons. Many of that generation can be found in our own body—in the sense organs of touch, muscle stretch, blood pressure, hearing, and balance. But many more are at other zoological rungs—in the lateral-line organs of fish (the precursors of our organs of hearing); in the gravity sensors of invertebrates; in the sensors of muscle stretch of lobsters, leeches, and worms; and so on.

We may perhaps also include here our "sense of temperature," although the corresponding sense organs and demons have proven rather elusive. The difficulty for the investigator here is that many mechano-sensory organs will respond to temperature unspecifically (their stationary digital-signal rate increases as the ambient temperature is raised over a considerable range). However, if the temperature sensors have independent status at all, their demons are likely to sense the stresses in their molecular structure caused by the temperature changes, so those sensors would rightly belong with the mechano-electric transducers.

But be that as it may, the photo-electric and mechano-electric transducer pigeonholes accommodate a vast expanse of sensory demons—and they do so naturally, as these are the very pigeonholes Evolution used for her selections.

A Lesson in Economics from a Master

Now to the signals. These are undeniably electrical, though this doesn't mean that they are like the signals of technological transducers—the photocells, photomultipliers, strain gauges, and microphones we gird ourselves with nowadays. The signals of those transducers are made of free electrons, whereas the signals of biological transducers are made of inorganic ions and water.

We would have a hard time finding an engineer using that sort of material for signals in our technology of communication. Yet it is the material of choice of Evolution. In her laboratory it is used to the hilt, and it's easy to see why. Not only is such liquid matter abundant, but given the legacy from single-cell times of electrochemical gradients across cell membranes, it is economically made into signals. Sodium ions, for instance, are 12 times more abundant outside than inside many animal cells and (free) calcium ions, about 100,000 times, and the cell interior is electrically negative with respect to the exterior at rest. So the corresponding channels need only open and the ions will rush in, giving a loud signal.

The way these signals are disseminated is no less economical. This is owed to the fatty component of the membrane, the lipid matrix. That matrix consists largely of phospholipids—fatty acids with a head group of phosphate or some other electrically charged group. Such molecules arrange themselves in two layers on the cell surface. That comes naturally to those lipid molecules. Their electrically charged head groups are

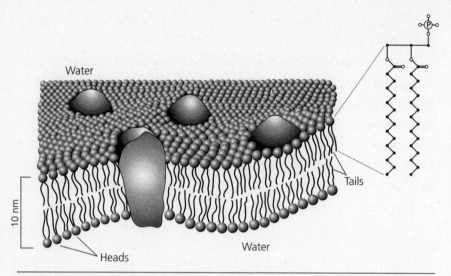

Figure 4.1. A Diagrammatic View of the Cell Membrane. It shows the lipid matrix, an ordered double layer of lipid molecules, with protein molecules embedded in it. *Right:* The chemical structure of a lipid molecule with its phosphate head group (*P*) and its two hydrocarbon tails.

strongly attracted to the water, so the molecules tend to orient themselves so that their heads will be in contact with the water, while their tails face one another (see figure 4.1).

This makes for a thin film—the two molecular layers are 7–9 nanometers thick, a two-thousandths of a human hair. Yet that film is tight. With about a million lipid molecules packed into a square micrometer, there is scant room between the fatty-acid tails. Strongly electrically charged atoms and molecules, like the inorganic ions and water, are kept out; their movement is slowed somewhere between a million to a billion times. The film, therefore, is a good electrical insulator, and it wraps around the entire cell.

The film also can store a fair amount of electrical charge (its capacitance is 1 μF/cm^2). Thus the charges the ions individually carry across are automatically summed up. Indeed, they will sum up to a sizable electrical signal. And this signal, without further ado, will spread over the film—the total electrical charge just distributes itself there, as it would over a capacitor in an ordinary electrical circuit.

Evolution made the most of these fine qualities in her design of electrical communication networks: first of all, she used the electrochemical gradients existing at the membrane (since single-cell times) to let the ions flow downhill; then she enlisted the electrical capacitance of the membrane to build up the informationally insignificant individual charges of these ions to full-fledged signals and spread them over the cell surface. And all three things—the downhill ion flow, the signal buildup, and the (passive) signal spread—at no extra information cost.

We have seen that the lady is thrifty, but here she rubbed the print off the dollar bill.

The Silent Partner

Much of the sensory operation thus rests on the lipid of the cell membrane. And I don't mean this in a trivial sense, like the lipid providing a housing or infrastructure for the sensory and channel demons. The informational operation itself, namely two essential components, signal integration and dissemination, rest on the lipid. The lipid is the silent partner of the demons, we might say.

Lipids are old players on the biological stage. Without them, without the cell-surface insulation they provide, a biological evolution as we know it could hardly have gotten off the ground. Their molecular double layer provided the essential barrier preventing the largely water-soluble biomolecular information accrued over aeons from drifting apart and undesirable outside molecular information from drifting in. It also provided the necessary condition for natural selection: it demarcated the boundaries of informational individuality, enabling the molecular systems of one cell to compete with those of another. In short, it provided a bubble wherein life could prosper and claw its way up.

That bubble had been in existence long before the sensing and the channel demons came on the scene of life. But once they did, they would enlist it in their information enterprise; with its high electrical capacitance and resistance, the bubble provided a suitable way to get their bits of information from one place to another on the sensory cells.

Take, for instance, the dendrite of a typical sense organ, the Pacinian corpuscle, which detects vibrations in our skin. Here the bubble is roughly cylindrical and the flow of electricity relatively simple. When a mechanical stimulus is applied to a spot on the dendrite, a number of ion channels will briefly open up there, giving rise to an electrical current that flows inward at the spot and outward elsewhere (see figure 4.2). It is as if a hole had been made in the bubble, a tiny hole through which electrically charged particles (inorganic ions) go, producing a primary electrical signal. This signal spreads over the bubble as the electrical charges distribute themselves passively over it. The signal doesn't get very far, but far enough (tenths of a millimeter) to elicit a secondary digital electrical signal at the dendrite. And this signal is transmitted to the brain.

How Electrical Sensory Signals Are Generated

To help see how the information flows here, let us follow the course of an information carrier, say a potassium ion, as it wends its way from the inside to the outside of the sensory dendrite. We will skip outright

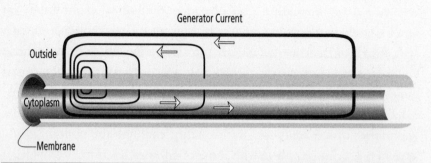

Figure 4.2. The Generation of an Electrical Signal in a Sensory Cell. The current flow (*generator current*) in response to an adequate stimulus. The diagram depicts the situation in a sensory dendrite, where a group of ion channels in the excited membrane locale (darker gray) has opened up in response to a mechanical stimulus, giving rise to an electrical current (*generator current*) flowing inward at the stimulus locale. (After Loewenstein 1959, 1960.)

the vagaries in the inside. There, in the cell water (cytoplasm), the ion's trajectory is inherently blurred by randomness. But inside the narrows of a membrane channel, where the movement is constrained by protein helices and their oxygen atoms, that whimsicalness gives way to more predictable behavior, which is what we will focus on.

Let's scale down our angle of vision to that of the single information carrier. What does a tiny ion see on the underside of the cell membrane? Far and wide, it sees a virtually impenetrable barrier that is occasionally interrupted by protein helices forming membrane channels. There, if a mechanical stimulus chances to hit that spot of membrane, the ion can slip through to the other side, and so can a number of other potassium ions in that membrane locale. Indeed, for the few milliseconds that the channel stays open, some thousand ions slip through there*—or better said, file through, because in the narrows there is but space for a single file. Each ion here carries one elementary electrical charge across the membrane, producing a change in voltage. That change would have no informational significance by itself—it is smaller than the prevailing electrical noise. But it adds up with the changes produced by the thousands of other ions filing through the channel. And that sum, in turn, adds up with that from neighboring channels, all this usually amounting to a few hundredths of a volt—a signal clearly standing out of the noise.

So much for signal integration. As for signal dissemination, the role of the lipid is just as laissez-faire; the signal spreads over the capacitive and resistive components of the membrane, falling off exponentially with the distance from the open channels (see appendix, item 6). It is a steep fall; a few tenths of a millimeter away, the signal already has dropped below the noise level. It's a much steeper drop than engineers are used to in their electrical transmission lines. But it is as much as one may hope for with lines made of lipid and protein. In any event, and this is the crucial point, the signal gets far enough to activate the voltage-sensing channels in the dendrite membrane, which will retransmit the

---

*Such a high ion throughput is typical for membrane channels. With the electrochemical gradients normally prevailing at cell membranes, the fluxes in potassium, sodium, calcium, and chloride channels are on the order of 10^6–10^7 ions/second/channel.

original sensory information in digital electrical form to the centers of the brain. And that is how much of the digital-signal traffic coursing to our sensory brain normally is generated.

―――――――

Let's take stock of the basic sensory operation. There are three participants in that operation: a protein molecule that picks up the outside information; another protein molecule that translates the information into an electrical signal; and a lipid partner, the molecular double layer of the sensory-cell membrane, which contributes the wherewithal for the integration and initial dissemination of the signal. The first two are the active participants, the cognitive heart of the operation. They contribute what in nonbiological sensors, like strain gauges, microphones, and photocells, is lacking: the Maxwellian demon element, the fundamental element of biological cognition. Both of the aforementioned demon types already existed in single-cell times, ministering to different and separate functions. But eventually they joined up in common enterprise. They started as equals, and even today each shows his independence in going through separate cognition cycles. But sometime during the long march of Evolution, one of them picked up a few extra bits (e.g., I_M) and got sway over the other.

Well, so it goes: all molecular demons are created equal, but some are more equal than others.

CHAPTER FIVE

Quantum Sensing

After transducers, demons, and George Orwell, consider our eyes.

Hooded above and pouched below, one is easily mesmerized by these two appendages of the brain. But it's what is deep inside, on the underside of the retina, that really counts: the demons of vision. These made the scene about 900 million years ago, some only 40 million years ago. They are Johnnies-come-lately by evolutionary standards, but they rose fast in the world. Their orb of influence grew from just a handful of neighboring channels to start with, and grew and grew . . . and as if to make a pitch for Orwell's words of wisdom closing the preceding chapter, their orb would in the end extend over the entire brain.

But what is it that makes the vision demons cast all others into the shade? What do they have that the others don't? They are astonishingly efficient as transducers—they make 1 picoampere, a macroscopic electrical current, from 1 elementary particle, a photon. But some demons of the mechano-electric kind are not so far behind. No, it's not just efficiency that gives the vision demons clout, but their quantum-sensing talent. They are capable of sensing a single quantum of light!

No number of exclamation marks could do such prowess justice—it borders on the fantastic. But this is precisely what sets these demons apart and makes them stand head and shoulders above the crowd. We have met by now quite a few protein demons, and all would make Maxwell's demon proud. But what they sense are molecules, particles incomparably larger than those the vision demons sense. The difference is not just quantitative but qualitative. We come upon a boundary in physics here: the borderland between the coarse-grained and the quantum realm—two worlds apart. We shall see more of this dichotomy and the cognitive chasm it represents when we deal with the mystery of our states of consciousness. Here, at the seedbed of those states, at the brain periphery, let's just say that the demons divide their interests accordingly: some have their antennas directed to the coarsegrained world and others to the quantum world.

The Quantum World

This is a strange world by any reckoning. Its particles, which make up everything in the universe, are minuscule—some of them have no mass at all. And the quantum world is full of such airy nothings: the photons, the elements that carry the electromagnetic force, and the mysterious gravitons moving with the same speed (the speed of light), which carry the gravitational force. The electromagnetic force causes the electrons to go around an atomic nucleus, much as the gravitational force causes the earth to go around the sun.

Or taking the information view, you may think of these quantum particles as communicating a message between nucleus and electrons, telling the electrons to go around the nucleus, or communicating a message between the sun and earth, telling the earth to go around the sun. That way of thinking would be equally valid.

The photons comprise a wide spectrum, ranging from the cosmic γ-rays, a hundredth of a nanometer long (about a tenth the size of a small atom), all the way down to radio waves, several miles long (see figure 5.1). The antennas of the vision demons are tuned to a narrow band, about a third of the way in between. The band begins with the violet (400 nanometers) and, progressing through the spectrum of blue, green, yellow, and orange, ends with the purple-red (720 nanometers).

Photons, like other quantum particles, have a ghostlike character. They can suddenly disappear and just as suddenly reappear. A photon, for example, visible as it comes from the sun to us, becomes invisible when it strikes an atom; its energy is used up for shifting an electron of the atom to a farther orbit. And it's not just a matter of becoming lost

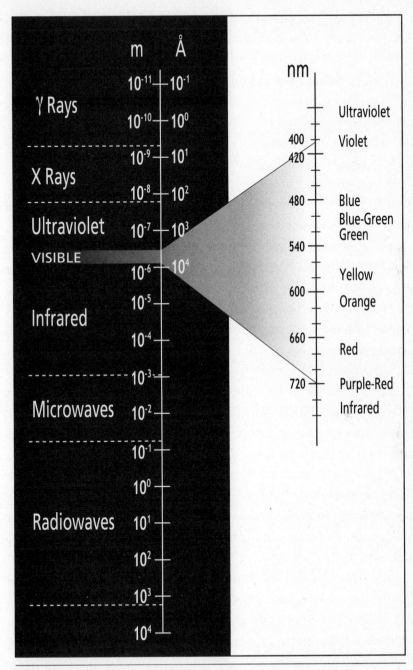

Figure 5.1. The Photon Spectrum. The electromagnetic wavelengths are given in meters (m) and angstroms (Å) on a logarithmic scale. Atomic radii are of the order of 10^{-10} m. *Right:* The visible waveband on an expanded linear scale in nanometers (nm). 1 nm = 10^{-9} m = 10 Å.

to sight. The photon particle actually changes character and converts into a *virtual photon*—a dematerialized one, as it were. And when the electron shifts back to a near orbit, the photon rematerializes and is given off by the atom as a visible particle.

Such quirky behavior is common fare in the quantum realm— whole hosts of particles can spring unbidden from the void. And that's not something easy to make friends with. Indeed, quantum theory was initially greeted with strong skepticism (one of the skeptics was Einstein), and a number of crucial points were in hot dispute. Niels Bohr, a pioneer of the field, summed it up when he quipped, "Anyone who isn't confused by quantum theory, doesn't really understand it."

Well, now, after 70 years, the storm has abated and the main points have gotten across—though a sense of wonder still lingers. And there is no helping it, as there is nothing in our daily experience that prepares us for the behavior of the quantum world—even its logical framework is staggeringly alien. But we now know why, and that goes a long way to opening one's mind to something that is so counterintuitive. In the past few years profound insights have been gained into that peculiar logic structure, and even though they don't yet fully fill the gap between the microscopic and the macroscopic, they have brought us closer than ever to a unified conception. Indeed, these days quantum theory has become so compelling that there is hardly a doubting Thomas left. Furthermore, the theory has had plenty of practical spin-offs: transistors, tunnel diodes, semiconductors in computer chips, lasers in compact disc players, magnetic resonance imaging in hospitals, superconductors—today an estimated 30 percent of the U.S. gross national product derives from quantum theory.

As for the now-you-see-it, now-you-don't aspect, that is but proper behavior in the quantum sphere; it's strictly according to Hoyle as far as quantum logic goes—and physically, perfectly legal because the momentum of the photon and the extremely short distance it moves during its conversions satisfy Heisenberg's uncertainty condition.

We will return to this further on where we tackle the workings of the cognitive brain. We will also explore then the physical conditions that enable photons and other quantum particles to exist in superposition and materialize in places that ordinary logic would deny them. Here, as a foretaste, let's consider the logic itself and size up part of its framework. This takes us straight to the bottom of what makes the quantum world seem so wondrously strange to us.

Think of what we do in our daily reasoning—when we put two and two together and come up with a conclusion. We may do so in various ways, but what they all ultimately come down to is an exchange of the language connectives *and* and *or*. These serve as our operators of logic, and together with *not* (the operator for negation), they constitute a complete set for all our reasonings. The mathematical logician George Boole cut through this in the nineteenth century. He showed that any logic or arithmetic task can be reduced to three operations: addition, copy, and negation; that is, any such task, no matter how complex, can be accomplished by a combination of the corresponding operators *and*, *or*, *not*.

That combination follows certain rules—Boole called them the laws of thought. Under those rules—specifically, under what became known as the "distributive law" of logic—the operators *and* and *or* can be exchanged. This is our commonsense modus operandi—it's what Hercule Poirot does when he lets loose with "*either* the butler did it *or* the maid did it," or what we do when we say that 2a + 2b equals 2(a + b). Such reasoning, however, would lead to utter nonsense if it were applied to the quantum world. Not that there is no logic in that world; *and*, *or*, *not* can serve as operators there, and do, but they must be combined differently, as they follow a different distributive law. In fact, quantum systems can perform any logic and arithmetic operation— and with stupendous speed.

Our Windows to the Quantum World

I have held that strange world at arm's length so far. While we dealt with demons who work with molecules, there was no need to go beyond the familiar world of coarse-grained things. But the vision demons have their feelers out in the other world, and what *they* grasp is in there. So now, whether we like it or not, we must cross the Rubicon.

It's not just the demons in our own eyes who beckon us to come hither. Vision demons are found high and low—even clams have eyes. Those demons are not all tuned to the same wavelengths ours are, but most, if not all, are quantum sensors. Thus the first issues before us are how the vision demons pull information out of the quantum world and how they make electricity from photons.

Physics and technology offer some guidance here, but only to some extent. There are devices galore in our technology—photocells, video cameras, image intensifiers, and so on—that make electricity from photons. A photocell made of a cesium plate, for example, will give off an electric current if you shine a light on it. The current gets generated as the photons strike the plate and knock electrons out of the metal surface. This conversion of light into electricity is a straight transduction: one photon comes in and one electron comes out—an energy tit for tat.

But there is not much cognitive finesse here, no Maxwellian selection. Any photon of light will do, provided it has enough energy; a photon with more energy will just make the electron bounce off faster.*

A vision demon, on the other hand, is selective—it is first and foremost a Maxwellian entity. It chooses among the visible photons, distinguishing a blue photon from a red one, or even from a green one. And for every photon it chooses—for every round of cognition—it pays the required thermodynamic price.

And here is how he does it. The demon traps the photon with the aid of a yellowish pigment nested in his center (see figure 5.2). This pigment, the *retinal*, has a chain of carbon atoms with alternating single and double bonds—a structure that brings the carotenes to mind, the primordial pigments that gave entrance to the First Information Arrow (see figure 2.4). Indeed, retinal is a close relative of those pigments; our body actually makes it from a carotene, a β-carotene called vitamin A. And like those ancient photon traps, retinal has delocalized electrons critically poised to catch photons of a certain wavelength. The retinal lies inside

---

*This phenomenon is the *photo-electric effect*, the physics platform on which Einstein, in one of his revolutionary papers in 1905, advanced the notion that light is made of particles.

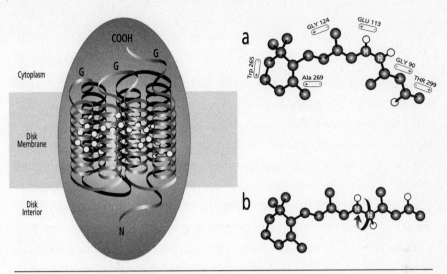

Figure 5.2. A Vision Demon. *Left*: Diagram of the demon's seven α-helices and the retinal at center. *COOH*, carboxyl end; *N*, amino end; *G*, docking sites for G-protein messenger. *Right*: The carbon chain of retinal (*a*) its twisted inactive form; (*b*) its untwisted form, after the capture of a photon; the arrow indicates the direction of the torque around the *11-12* bond. Carbon atoms, black; hydrogen atoms, white. The position and amino-acid sequence number of some neighboring dipoles instrumental in the tuning shift have been sketched in.

the protein bulk of the vision demon, the protein *rhodopsin*, and is tuned to the visible wavelengths. In the dark, the retinal carbon chain is twisted around a strategically located double bond (bond *11-12*); when it catches a photon, it straightens out—the energy quantum absorbed converts to a torque, untwisting that bond (see figure 5.2b).

What we have here is a photo-mechanical transduction; the absorption of the photon quantum causes the retinal carbon chain to move. But this is only the first step. In the next, the movement of the chain is transmitted to the demon's protein structure. The end of the retinal chain is linked to one of the protein's α-helices, and a hydrogen nucleus can move across here. So, all in all, the photon information flows smoothly from sensor to protein, and the protein gets activated, responding with a configurational change of its own. It is in that configuration that the demon finally interacts with his channel partner, by sending the crucial I_M command.

Coherent Quantum Information Transmission

That, in a nutshell, is how things go at the molecular level—the bio-chemical picture. But the greatest wonder, and profound secret, stands revealed when one gazes deep into the atom with techniques of quantum physics. Then, as one probes the retinal carbon chain, events unreel with lightning speed. And it's literally lightning. Upon photon absorption, the retinal carbon chain instantaneously experiences a torque about the strategic bond, and the primary configurational change of the protein—a transitional state biochemists call bathorhodopsin—comes close on its heels. The whole process, including the primary configurational protein change, is over in just 200 femtoseconds (0.2 trillionths of a second); what is more, the efficiency with which energy and information are transferred border on the extreme (overall quantum yields of 0.7).

That alone speaks volumes. Such transmission speeds and efficien-cies are unheard of in ordinary chemical reactions—or anywhere else in the macroscopic sphere. They are known to occur only in the micro-scopic sphere, the realm of ghostlike quantum particles.

Indeed, we might say they are the signature of those particles. Quantum particles can behave like particles or like waves. All elemen-tary particles can behave that way, including small atomic nuclei like the hydrogen nucleus (see box 5.1). Take the photons, for example. As particles, a stream of them will knock out electrons from the surface of a metal (photoelectricity); as waves, they will diffract or produce inter-ference patterns, sequences of light and dark bands caused by waves arriving successively in step or out of step. When they are in step they will transmit energy and information with extreme efficiency, as their peaks and troughs then add on one another. It's a bit like ocean waves in lockstep. Physicists have a catchword for that sort of wave behavior: *coherence*.

Now coherence, namely quantum-wave coherence, is precisely what is found in photoreceptors. This pivotal piece of knowledge is relatively new—we owe it to the quantum physicist Charles Shank and the physical chemist Richard Mathies and their colleagues. They managed to probe

Box 5.1. Quantum Waves

The concept of quantum waves stems from a proposal in the early 1920s by a graduate student of physics at Sorbonne University, Prince Louis de Broglie. Reasoning that Einstein's $E = mc^2$ relates mass to energy and that Planck's equation relates energy to the frequency of waves, mass should have a wavelike nature. So, in his doctoral thesis he put forth the notion that a moving electron should have a wave associated with it. His examiners at the Sorbonne found the idea that particles are also waves too fantastic to be taken seriously, though they accepted his thesis defense as a virtuoso performance. But his thesis adviser, Paul Langevin, was not given to passivity; he spoke to Einstein by telephone about the matter and arranged that an extra copy of de Broglie's thesis be sent to him. He didn't have to wait very long; Einstein wrote back, "Er hat eine Ecke des grossen Schleiers gelüftet" ("He has lifted a corner of the great veil").

Thus started wave mechanics, a vigorous branch of physics, which within a few years, through Erwin Schrödinger's mathematics, would acquire enormous explanatory and predictive power. Schrödinger's equation describes in probabilistic terms the evolution of the wave, the so-called *wave function,* Ψ (pronounced "psi"), which gives all the possible alternatives that are open to a particle system. One of the great achievements of twentieth-century physics, this equation touches nearly all aspects of matter: the structure of the atom and nucleus, the mechanical properties of solids, the conduction of electricity, and the chemical bonds.

Today the notion that matter—all fundamental particles of matter, including the quarks—has a wavelike character has firmly taken hold. Much of modern physics is built on it, and on the practical side it yielded the transistor, the superconductor, television, and so on and on.

But why then do we experience matter as something so stiff and sturdy, and not as waves? De Broglie's equation provides a simple answer: the wavelengths are extremely small; they are proportional to Planck's constant, a minuscule number. Matter waves are therefore unnoticeable in our day-to-day world.

the retinal-bathorhodopsin system with high-resolution spectroscopic and nuclear-magnetic resonance techniques. Using very short light pulses for their probings (on the order of 10 femtoseconds), they followed the motion of atomic nuclei in bathorhodopsin and showed that they behave like coherent quantum waves (the coherence lasts throughout the primary chemical reaction). And rounding out their experiments, they manipulated those nuclei and showed that the information-transmission efficiency takes a nosedive when the nuclear motion is slowed.

Thus, by no stretch of imagination is this run-of-the-mill biochemistry or molecular biology. Quantum mechanics, not classical mechanics, rules the roost at this sensory outpost of our brain.

And one more thought. Insofar as the core of the sensory system here belongs to the quantum realm, it may be expected to follow the strange rules of logic I mentioned before. And it does. But things run even deeper. Not only does it abide by those rules, but the tiny retinal and its protein partner, as we shall see in a later chapter, actually have the wherewithal to implement quantum logic and to compute.

The Advantages of Being Stable

What happens in the rest of the rhodopsin structure is more prosaic. The information transmission proceeds at speeds of macroscopic mechanics here, and there is no reason to suspect that quantum mechanics are involved. But there is something that may make you wonder. How does the information from a single photon make it all the way through the protein structure? How does it overcome the noise? The rhodopsin protein is 348 amino-acid units long, a conglomerate of seven α-helices weaving in and out of the membrane. So how does such a hulk give an informationally significant response to *one* energy quantum, a response that stands out from the background noise? Thermal noise is a perpetual nuisance in biological information transmissions, because the molecular structures that carry the information get jostled about by the thermals. So all the quantum sensing would do no good if, by that jostling, rhodopsin would go into the active configuration and send out an I_M command.

Well, we needn't worry on that account. Spontaneous rhodopsin activation happens—there is no helping it at ordinary temperatures. But it happens very, very rarely: once in 800 years. Evolution has seen to that. Over aeons of selections, she eventually seized on a protein that has an extremely low probability of becoming activated in the dark—a demon who is locked into the inactive configuration, and only a photon of the right wavelength stands a reasonable chance to unlock it.

Thus the secret of quantum sensing lies as much in the molecular stability of the vision demons as in their talent for catching photons and transmitting their quantum information coherently. All the rhodopsins—and this includes those in mammals, birds, reptiles, amphibians, fish, cephalopods, and insects—are superb photon traps, but they also are superbly stable in the dark. It is this rare combination that makes them the prodigies they are.

Why We See the Rainbow

But we are aware by now, if there ever was any doubt, that the vision demons are the cream of the crop. The three types in our eyes collectively cover the visible photon spectrum, a waveband extending from azure blue all the way to crimson red (figure 5.1).

But how do these rhodopsins manage to sense all that? It's not only that the waveband is so wide, but it goes far beyond what their primary sensory component, the retinal, can sense by itself. This question goes to the heart of the problem of how we see colors—an old puzzler, but besides quondam perplexities, there are new ones. Isaac Newton opened the subject 300 years ago when he found that white sunlight splits up into a rainbow as it gets refracted. But why we see it in all its glorious polychrome, we have only begun to understand in recent years through refined probings of rhodopsin's inner structure.

A trenchant and singularly fruitful hypothesis was offered as early as 1802 by Thomas Young. An advocate of the wave theory of light, Young ascribed our color sensations to three types of receptors in the retina, which are attuned to different wavelengths. Young was the proverbial Renaissance man—physicist, physician, and Egyptologist. As a physicist, he knew that when blue, green, and red light (the "primary colors") are mixed in equal parts, they yield white, and that mixing those primaries in various proportions—something painters have down pat—yields multiple hues. As a physician, Young was aware of various forms of color blindness, especially the one afflicting his contemporary, the physicist John Dalton, who was unable to see red or red-to-green hues. On these grounds, Young put forth his hypothesis of three distinct receptors, three sensors resonating with different wavelengths. In his words: "As it is almost impossible to conceive each sensitive point of the retina to contain an infinite number of particles, each capable of vibrating in perfect unison with every possible undulation, it becomes necessary to suppose the number is limited, for instance, to the three primary colors."*

Young's idea took over a century to receive the consensual nod, but it could hardly have been more prescient. His three "particles" eventually turned out to be three proteins, the three rhodopsin types inside our sensory cells of color. There are three kinds of such cells in our retina, the *cones*, each with its peculiar rhodopsin type tuned to a certain waveband. These cells convey the photon information to the brain. They are about 2.5 nanometer in diameter (figure 5.3), so any image in our eyes—even as small an image as that of a point source of light—involves a considerable number of cones, and the relative extent of their excitation somehow determines our sensation of color. From the cones the photon information flows to the second- and third-order neurons of the retina (the *bipolar cells* and the *ganglion cells*). These are the first information processing station—our brain's palette, as it were.

---

*Vision was a popular subject among physicists at that time. Only eight years before Young put forth his hypothesis, John Dalton, the father of modern atomic theory, had presented a paper on the color blindness that afflicted him to the Manchester Literary and Philosophical Society (1794). "That part of the image which others call red," he wrote, "appears to me little more than a shade or defect of light." Maxwell later (1855) did a systematic study of Daltonism and concluded that those affected lacked either the green or the red sensor. Eventually, after neurophysiology got its cellular footing (the botanist Matthias Schleiden proposed his groundbreaking Cell Theory in 1838), Hermann von Helmholtz (1867) applied Young's notion to the cellular units of vision and proposed that our color sensations result from the relative extent of excitation of three distinct sensory cells.

But we left dangling the most basic question, why the sensitivity of the rhodopsins ranges from the blue to the red, when that of retinal itself goes only as far as the blue. I will put this in terms of the wavelengths that the pertinent electrons absorb, so that we may see exactly what we are up against. The maximum photon absorption of the (protonated) retinal isolated from the protein is at a wavelength of 440 nanometers, whereas the maxima of the three rhodopsins are at 425, 530, and 560 nanometers. Clearly we are dealing with a shift in photon-trapping wavelength here—a tuning shift. So if at first blush the thought of optical filtering should crop up, it is no go. No amount of filtering by the protein could give rise to a tuning shift. Such a shift implies a change in organization of retinal's electron cloud, a change in the quantum state of electron orbits. But how?

One could think of several ways that a change in electron quantum state might come about, but with proteins, the most likely way is through an interaction with electrically charged components of the proteins. Rhodopsin has a number of such components, amino acids carrying a positive charge at one site and a negative charge at another—or *dipoles* as such polarized molecules are called for short. These amino-acid dipoles are close enough to the retinal to fill the bill. For example, rhodopsin's amino-acid unit 113, a glutamic-acid unit, is ideally positioned to influence the ground-state electron distribution of retinal's carbon chain (figure 5.2). Indeed, when one experimentally replaces this unit with an uncharged amino acid (glutamine), the absorption maximum of rhodopsin veers to the ultraviolet, the top of retinal's range.

If we survey retinal's nesting place in the protein, we find about a dozen such amino-acid dipoles there, and they are all in a position to affect retinal's electron cloud. Recent studies combining experimental mutations with resonance Raman spectroscopy, in fact, bear this out. They show that these dipoles make themselves felt on the electron charge distribution of the photon-excited retinal, as well as that of its ground state. In general, the charged amino acids neighboring retinal's carbon chain (like glutamic acid 113) predominantly act upon the ground state, whereas those near retinal's ring act upon the photon-excited state, dragging rhodopsin's tuning toward the longer photon wavelengths.

And this is why we see the rainbow—amazing what a few electrical charges in the right place will do!

The Demons Behind Our Pictures in the Mind:
A Darwinistic Physics View

But how did those dipoles come into being? What evolutionary pressures might have molded the rhodopsin molecules and the fateful matrix of their charged amino acids? This may seem an unconventional way to ask about the origin of our eyesight, but it puts the finger right on the spot. Besides, it may provide us with a lead to that arcanum arcanorum we call pictures in the mind.

But before anything else, we must define what we mean by *color*. Though it is one of our most common perceptions, color is not something easily explained. Try to define, or even describe, a color, and you discover how quickly you run into tautology. One naturally looks to physics in this case. But if you ask a physicist why yellow is yellow, it won't get you very far. For him, it's just electromagnetic waves about 590 nanometers in length; the fact that one can match it by mixing waves of 760 nanometers (which by themselves produce the sensation of red) with waves of 535 nanometers (which by themselves produce the sensation of green) baffles him as much as anyone not in the know about the rhodopsin demons' dipole cabal. There is no physics reason why two spots, one lit by a single wavelength and the other by a mixture, should look exactly alike. No physics theory will predict that.

An experienced painter does better—or so it would appear, as she foresees the outcome of her color mixing and makes few mistakes. But it won't take much to see that all she does is work by rote. No explanation here, no prediction in a scientific sense. Granted, there are color charts and the so-called color vectors and color triangle, which art and other textbooks make noise about. But those are just empirical props.

So, caught between the devil and the deep-blue sea, we'll try a little derring-do. We pose the color question in evolutionary terms, though always with a view to physics. Building on what we know about the rhodopsins' amino-acid dipoles, we ask what natural selection pressures may have caused them to be deployed the way they are. We will take our cues from the solar electromagnetic spectrum.

To begin with, let's put the color question in a broader context. The physicist is right, of course; color is not a thing of the world outside but of the one inside us. It belongs with the other aspects of our conscious experience, like flavor, scent, and tonality; and the fact that the last goes by the name of Klangfarbe (sound color) among musicians underlines the kinship. Those are the *qualia* of psychologists, which color the pictures in our mind. Such pictures are representations of the outside world, which our brain renders from many and various pieces of information that come in through the senses—in the case at hand, largely solar-photon information. The photons that the sun showers onto our planet come in all sorts of energy gradations and so serve to supply information about spatial dimensions and energy states of objects, as they bounce off the objects' surfaces into our eyes. The photons peel information off the objects, as it were, and they do so in great detail and with speed.

All that horn of plenty is free for the asking—provided you have the right demons, of course. If you do, you have an obvious advantage over competitors who don't: your world picture will have a higher time resolution. Such striving for temporal detail, we may imagine, is the sort of selection pressure that, late in multicell organismic evolution, may have driven brain development toward the sensing of photons—it's a late echo of the pressures for sensing information transients we saw in chapter 3. While before, the brain's world pictures were largely made from information coming from ponderous water-dissolved or airborne molecules (chemo-sensing), now they would take wing on photons.

And what a wing it was that would one day allow the lion to chase the gazelle across the grassland or humans to aim an arrow at the flying grouse. The power behind all that is the rhodopsin demons. These are not only chief actors in the vision process, but in a sense, the prime movers of the related brain development. They are the products of Evolution's perpetual search for molecules that would match the photon quantum—and in this belated instance, a search for molecules that would match a diversity of quanta. This strategy centered on the discreteness of solar photons and, putting their natural diversity to good use, it would eventually yield the informational material from which a developing brain would construct a world picture with unprecedented temporal and spatial detail. We know that the search came up with protein-retinal conjugates that expanded retinal's native photon-matching capability, tuning it to a broader band of photon wavelengths.

Now, against that backdrop, we will play the physicist—or perhaps more aptly put, the Darwinistic physicist. We ask where, on physics grounds, those wavelengths were likely to lie. The answer is plain as plain can be: between 720 and 400 nanometers; that's where the photons coming to Earth were, and still are, most abundant. And this is precisely where the rhodopsins have tuned the retinal to.

Next we take a stab at a sharper question: Where, within that spectral region, was the highest sensitivity likely to be, on physics grounds? Again the answer comes naturally: where the solar radiation is strongest, at 590 nanometers, which is the color yellow—exactly what we are most sensitive to.

Why White Is White

That was, as the Michelin Guide would say, worth the detour. So let's go on and see what else about color we can retrieve from the heap of chance.

We start with white. That is a relatively easy target from the vantage point of our 720–400 nanometer cynosure. We can define it as the condition wherein all wavelengths in that band are equally represented—an even distribution. Most other colors are not so forbearing. They are mixtures—though keep in mind that they are information mixtures, not wave mixtures. The actual mixing happens after the wavelengths have been sorted out and the information has left the wave carriers. It is the wave information in molecular form that gets mixed, and this takes place in a cellular space beyond the sensory cells, the second- and third-order retina neurons (see figure 5.3).

These neurons receive information coming from the three types of cone cells (where the original wave information was broken down). Here that information gets integrated. It is the site of a triple conflux—

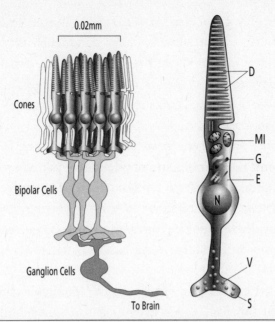

Figure 5.3. Our Daylight-vision Cells. *Left:* A view of a region of the retina displaying a group of cone cells and their neuronal connections. The ganglion cell collects the photon information from this group, and its axon conducts the information to the brain; conjointly, such axons constitute the optic nerve of an eye. The calibration bar is about the width of a human hair. *Right:* Anatomy of a cone cell. (*D*) the stack of disks; the rhodopsin demons are embedded in the disk membranes (see figure 5.2). (*N*) nucleus, (*G*) Golgi apparatus, (*E*) endoplasmic reticulum—the organelles instrumental in making the molecules involved in the sensory cell function—(*MI*) the mitochondria that make the ATP. (*S*) synaptic terminal. (*V*) vesicles containing the molecular information to be transmitted to the bipolar cell—the second-order neuron.

the "palette" I mentioned before. That metaphor implied that there is a choice in the mixing. Indeed, these cells possess a set of demons who select a part of the incoming information, while the rest is quashed.

The question now is, what are the standards for that selection? And there's the rub; they are utterly ad hoc. Those standards go way back, to the time in evolution when the vision circuitry of the brain emerged, and like the rest of the nascent brain, that circuitry was geared primarily to survival. So what those demons select is information useful to this selfish end—and nothing else gets past the second- and third-order neuronal filter. The primal brain doesn't give a hoot about exact wave-length; it's contrasts, contours, and, at a finer level, energy states of atoms and molecules that it really is after—anything that serves to tell one object from another.

Small wonder the physics of color seems so unruly; color goes by historical, not absolute, rules. And compounding the problem, the rules are written for reflected lightwaves, rather than direct ones. Indeed, if one isn't in the know about the information conflux past the wave sensors, one may well think that our perception of color somehow is perversely determined by what we don't see.

Well, that really wouldn't be too far off the mark. The ancestral logic of the brain is a bit like that. It is a logic concerned with differential information.

Consider the color of this page and the color of a blue shirt. The background of this page is white, because its paper molecules do not much absorb wavelengths other than the infrared and ultraviolet. The rest, meaning the whole 720–400 nanometer band, gets reflected to your eyes, producing a color sensation. And because these reflected wavelengths have roughly the same distribution as the ones originally hitting the paper, the sensation is white—all the wavelengths are equally represented (see the definition above). On the other hand, the molecules of the blue shirt absorb waves over a broad range, namely, those in the long-wavelength regions (red) and mid-wavelength regions (yellow and green), whereas only waves shorter than 520 nanometers get reflected into your eyes, namely, wavelengths corresponding to blue hues.

It is thus the difference, what's left over from the 720–400 wave band, that counts. However, the subtractions can get more elaborate when the leftover wavelengths are scattered over the band. Take, for instance, the ink I used for writing this manuscript. That ink contained copper molecules that strongly absorbed the waves around 620 nanometers, while the rest of the band, a whole slew of sundry wavelengths, got reflected. That combination of wavelengths produces a blue-green sensation.

The perception of that hue is not uniquely determined—other wavelength combinations produce a blue-green sensation, too. However, it's not something arbitrary—the hue will invariably ensue under the same conditions—it's just that our brain uses an ambiguous code. Such ambiguities are also found in other sensory systems. They are inherent in the information-processing hardware of our brain and tinge all of our world picture.

So unavoidably, our world picture has its twilight zones, and it's good to be aware from the start that they already begin at the brain periphery, often within earshot of the sensory receptors.

The Quantum View

Now, if we take the quantum view, things get stripped down to their essentials and the color question becomes a matter of photons and atoms. Both particles have definite energies. Photons carry energies in certain amounts—or one might well say, idiosyncratic amounts, for each particle in the photon spectrum has a particular energy quantum (inversely proportional to its wavelength). Atoms, on the other hand, carry energies in their electron clouds; those, too, are idiosyncratic, as the energy of an electron is restricted to certain values. Electrons are popularly pictured as orbiting around an atom's nucleus, like the planets around the sun—and for many intents, that is not a bad metaphor. But when it comes to energy, the facts won't bear it out. Atoms can exist only in stable configurations where the electrons are confined to certain orbits—a discrete set of orbits with fixed energy amounts. On the other hand, planets could in principle be moved farther away from the sun, or closer, by giving them any energy amount.

The fuzzy electron clouds around atomic nuclei, we shall see, are better described by Erwin Schrödinger's wave mechanics, which takes into account the wave-particle duality and statistical behavior of electrons. The differential equations here yield *wave functions* that specify the electron trajectories statistically—they give the odds of where the electrons may be found in some point in space (and the three-dimensional end result is a cloud of electron probability around the nucleus). The wave functions offer information about the probabilities of where the electrons may be found in space—indeed, they contain *all* the information that *can be* obtained about those particles within the bounds of the uncertainties in the quantum sphere. And within those bounds, they specify the path of an electron around the nucleus, namely, its *orbital* (not orbit, to betoken its probabilistic character, as distinct from the deterministic one of the paths of planets).

In general, electrons tend to occupy the lowest-energy orbitals in an atom, and when they do, the atoms are stable—a condition that goes under the name "minimum energy state" or "ground state." That state may be levered up with extrinsic energy, and photons provide a convenient source for that. A photon here will boost an electron to a higher-energy orbital. The energy requirements for that depend on the atom's orbitals. Many atoms and molecules require photons with a relatively high energy quantum—on the order of that of ultraviolet light, or higher. But there also are plenty of them around for which visible photons will do. And those are the ones that, in that curious differential information game we saw, provide the grain of our world pictures.

The boosting of electrons, however, is not a free and easy play. The photon's energy quantum must exactly match the energy difference between the orbitals here. Both photon and electron orbitals are quantized, so the photon quantum must equal the difference between the atom's ground state and the higher energy state—that's what the law of conservation of energy demands. But all the same, given the huge variety of solar photons, it is not difficult to get a match.

Again, Why White Is White

Let's see how the game is played and how it yields colors. Take blue-green, from our ink example. The copper atoms here have a relatively simple electron configuration, and to work out the orbital energies is a manageable thing. In any event, for the purposes here, all we need to know is the difference between the ground state and the next higher state. That difference works out to be 2 electron volts—a quantum that falls smack in the middle of the visible photon range: the 620-nanometer wavelength.

Now we see why, out of the entire spectrum, only this wavelength got absorbed. It's just that one type of ball that is up to par. The rest bounce off—and those balls are the ones that feast our eyes.

Although statistical, the game is methodical, even fastidious, and its outcome, predictable. The quantal increment of energy states, the sharp transition in an atom's electron cloud, determines, by subtraction, the colors we see. It's a statistical determinacy—not the sort of thing scientists usually like—but good enough for our brain. Indeed, from the cradle to the grave, we stake our lives on it. But just in case any doubts about the causality itself should linger, those, too, can be allayed: the same pattern of atomic energies will ensue if one experimentally supplies the necessary energy quantum to the ink's copper atoms with a beam of electrons, instead of photons.

So much for the blue-green. As for white, the quantum answer is just as simple. Here, too, the energy difference between electron orbitals holds the key, but the difference is too large to be matched by any photon between the ultraviolet and infrared—the orbitals are way too far apart. And that is what happens in the case of the paper atoms of our example; all of the photons there, for want of suitably spaced orbitals, bounced off, engendering that equable photon demography we call white.

Lady Evolution's Quantum Game Plan

With this bit of physics in the bag, we once more venture out on the evolutionary path. We now have a few fresh guideposts. Though these may not be enough to descry the visual brain development in detail, they may be enough to allow us to get a sense of the general strategy behind it.

What might that strategy have been? What was Evolution really after when she brought forth the vision demons and their neuron retinue? Well, it is easier to say what she was *not* after: a faithful representation of the outside world. As is evident from the ambiguities in color-coding, she was not interested in facsimiles of the world, but in a discriminating picture useful to the organism in the struggles of life.

Our sensory brain, thus, is blatantly pragmatic. And that's not just a quirk of its visual sector, but a feature of our entire sensory brain. Ever since the 1930s, when the great neurophysiologist Edgar Adrian tapped some of the sensory information lines in a frog, physiologists have been probing the lines in all sorts of organisms, including humans, and the results leave no doubt what kind of information serves as the primary database for our perceptions: *discriminative information, that is, information allowing the organism to tell one object apart from another or one environmental condition from another.*

And what could be better for that than the quantum states of atoms! These states are distinctive features of atoms. They are reliable identity badges—atomic ID cards as good as they come. And they get thumbed through for us by the photons in our environment at no cost. Those genial particles scan the world for us in passing, as it were, and even throw a rainbow or two into the bargain.

Thus, we can glimpse the outline of an evolutionary strategy. It's a strategy aimed at the physics bottom: *Evolution fixed on the existing parity between the quantum states of atoms and the quanta of solar photons, and betaking herself of two of the most commonplace energy fields on earth (the fields of electrons and photons), turned their interactions to her advantage.* Or look at it this way, Evolution's way: there were two overlapping energy fields in our environmental space, an electron field whose quanta were electrons and an electromagnetic field whose quanta were photons, and when and where those quanta matched, there was information to be had for nothing. And ever the scrooge, that is what she went after.

Quantum Particles That Don't Cut the Mustard

We have been focusing on photons here. This is only natural. These particles provide us with much of the information about the outside world—and every creature on Earth with eyes, from the dim-sighted scallop to the sharp-eyed eagle, uses photons. But what of other quantum particles? Our environment is certainly alive with photons, but they are not the only particles that the colossal nuclear reactor in our sky sends down to Earth; there are also muons, tauons, (free) electrons, and neutrinos. Could any of those have been used for sensory purposes?

Well, you needn't worry. I have no intention of stirring a tempest in a very arcane sort of teacup. We can give short shrift to those quantum types, if we place them in the historical context of brain evolution. The muons and tauons were never within easy reach in the earthly environment. Nor were the electrons in free form one or one and a half billion years ago when eyes evolved. Perhaps elsewhere in the universe, though, they might have been. This gives some food for thought regarding the possibility of extraterrestrial life; all three of those quantum types are electrically charged (−1) and hence capable of transferring information to molecules—and it boggles the mind what kind of world representation those quantum types might have wrought.

I leave that subject to the science fiction writers. The three particle types mentioned are of little concern to us; they were not contenders against the photons, at least not here on Earth. All we need to consider then are the neutrinos. These quantum particles then, as now, certainly abounded. Hundreds of billions of them come down to us from the sun every second. And they do so day *and* night, as they hardly ever interact with earthly matter. They have no electric charge and so do not respond to the electromagnetic force. Nor do they respond to the strong-nuclear force. The probability of their interacting with earthly matter, therefore, is extremely low. Or as John Updike once put it,

> *Neutrinos, they are very small*
> *They have no charge and have no mass*
> *And do not interact at all.*
> *The earth is just a silly ball*
> *To them, through which they simply pass*
> *Like dust maids down a drafty hall*

For sensing purposes the neutrinos are useless; being chargeless, they cannot transmit information by electromagnetic force. The only way they conceivably could transmit information is by the weak nuclear force, but that force is so weak and would require so deep an intimacy that the odds of information transfer are exceedingly small.*

In short, the photons were a Hobson's choice.

---

*The odds, however, are not zero. You can actually witness that rare event in our environment of a neutrino transferring information to an atomic nucleus—though you need to have the patience of Job. The high-energy solar neutrinos (14MeV) will transmute chlorine nuclei into argon. That is how physicists measure the flux of these elusive particles (they use chlorine bound in the form of C_2Cl_4, a cleaning fluid, or heavy water, whose deuterium nuclei are split up into their component protons and neutrons). Updike, in the third line of his verse, went a bit over the top. But poets are entitled to a little license; we scientists are not.

CHAPTER SIX

Quantum into
Molecular Information

Boosting the Quantum

In the preceding chapter we tracked the spoor of the photon information inside the rhodopsin. It is in the innards of this formidable molecular demon that the quantum information somehow crosses the threshold into the macroscopic world. But how does that minuscule amount of information make it to the centers of the brain? How does it even make it the few micrometers away through the sensory cell to the demon's channel partner?

We have seen before how this happens in broad outline—figure 3.4 gives the general scheme. However, we glossed over the noise problem there, and in communication-systems operating with free-floating molecular signals, the problem is most severe. Recall that the demon and his partner are in different parts of the sensory cell—one is imbedded in a disk membrane and the other in the surface membrane, with a gulf of salty water between them (figure 5.3, right). There is not a snowball's chance in hell that the information quantum could survive in that noisy gulf, unless it gets boosted up.

Indeed, it is boosted all along its way to the channels by a set of three specialized molecules forming a transmission chain, the I_M message chain of our general sensory scheme. Two of the molecules here are proteins. I_{m1}, the first message bearer in the chain, is on its way as soon as a photon is trapped; it leaves the rhodopsin demon in less than a thousandth of a second after a photon hit. It then diffuses laterally in the liquid phase of the disk membrane and passes the message on to I_{m2}, the next protein in line. And this passes it across the disk membrane to the cell water, where I_{m3} picks it up and carries it to the channel in the cell surface membrane. I_{m3} is a water soluble molecule, a nucleotide called cyclic GMP. All three molecules, and most vigorously I_{m2}, boost the information by dint of their sheer molecular multiplicity.

It is, in fact, an enormous boost, and we get a feel for it by fixing on what goes into the information chain at the beginning and what comes out of it at the end. We are in the position to do so thanks to the elegant experimental work of the neurophysiologist Denis Baylor and his colleagues. These investigators isolated sensory cells (rods) from the macaque eye and managed to measure the electrical signal put out by these cells, under controlled illumination. They showed that *the information from a single photon will change the open state of 200 channels, producing in the aggregate a signal of 1 picoampere.*

Everything here is geared to getting the photon's information quantum through the sensory cell despite the prevailing noise: the disk membranes are filled to the brim with rhodopsin molecules—there are 20,000 of them per square nanometer, the highest sensory-demon density anywhere, and the amplification provided by the members of the I_M chain is calibrated to compensate, with some margin to spare, for the dilution of the molecular information in the corresponding membrane and cytoplasm compartments. Occasionally the signal here may teeter-totter, but there is enough safety margin built into the system so that the signal rarely veers toward a noise level too close for comfort.

However, all that fine information-transmission machinery of the sensory cell would be spinning its wheels were it not for a special evolutionary adaptation keeping the noise low in the channels. The bore of the channels in these sensory cells is much smaller than that of membrane channels elsewhere—the ion throughput is but a thousandth of that found in membrane channels elsewhere—which effectively reduces the noise, owing to the statistical fluctuations in the number of opening and closing channels. And that is what brings the transmission of the information of the photon quantum inside the tortuous sensory cell to a happy end.

Here are the final figures: in the dark, the noise amounts to 0.3 pico-ampere, while the signal produced by a quantum of light measures 0.7 picoampere (the 1-picoampere value I gave above was a rounded-off figure). The noise thus amounts to about half the signal—just about right to see the signal safely through the cell.*

So what started as a quantum-coherent motion of a few atomic nuclei in the cell's retinal in response to a single photon ends up as a substantial electrical current. Precisely where and how the information quantum crosses the threshold into the macroscopic world we don't know—that piece of physics Lady Evolution holds close to her chest. But at least she lets us in on her information boostings on the macroscopic side—as if to show us by gentle grace that there are other ways of overcoming the noise than shouting.

A Consummate Sleight of Hand

Let's now take stock. We have a total of five distinct molecular demons participating in the game of photon-information transmission in our sensory cells of vision—and I include here the capture of the photon information. Two of them, the rhodopsin and the channel demons, are rather latecomers on the evolutionary scene—the very latest, one of the rhodopsins, arrived only about 40 million years ago. The rest of that demon fraternity were there long before brains were around and got co-opted from ordinary body cells.

---

*Another significant noise factor is thermal rhodopsin activation. Though rare (see chapter 3), such activation is not a negligible noise source in the dark. With 10^8 rhodopsin molecules in a sensory cell, such as our rods, it will give rise to about one discrete event every 2.5 minutes. Interestingly, this low-frequency noise gets through to our cortex. On the other hand, the noise owing to the channel fluctuations, which is of a much higher frequency, doesn't get to the cortex; it is filtered out at the first synaptic stations in the retina.

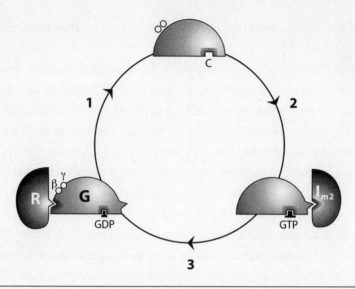

Figure 6.1. The G-demon Cycle. The diagram shows the demon (*G*) interacting successively with the receptor (*R*) and I_{m2} in the course of a three-state cycle: (*1*) After receiving the information from the active receptor, the demon switches to a conformation with an open cavity (*C*) and picks up a guanine trisphosphate (*GTP*). (*2*) He docks at I_{m2} and transfers the information, while splitting off one of the tetrahedra of the phosphate. (*3*) He switches to a conformation where the cavity is inaccessible to guanine trisphosphate and returns with the remaining two phosphate tetrahedra (guanine diphosphate, *GDP*) to the receptor to restart the cycle. In this generalized scheme, *R* stands for a rhodopsin, an odor or a hormonal receptor, and I_{m2}, for the corresponding cyclic-nucleotide phosphodiesterase. *β* and *γ* are subunits of the demon's protein that are necessary for his interaction with the receptor; they dissociate from the main body of the protein in state *1*. (After Loewenstein 2000.)

One member of that fraternity, $I_{m1,}$ deserves a closer look (see figure 6.1). He belongs to an ancient family, the *G-proteins*. These made their debut sometime between two and three billion years ago, still during the single-cell age. But they were stars already then and have been so ever since. It's easy to see why: they are amazingly self-sufficient in information terms. They are amazing even by demon standards; they perform full-fledged cognitions without an external source of information!

Not that they get a special dispensation from the Second Law—nobody does. They must balance the entropy books with external information (negative entropy), as decreed by that law. And they do so strictly according to the cognition equation (see chapter 2). But they need no extra help for that; they can make do by themselves, because they carry the required external information piggyback in the form of an organic phosphate, a guanine triphosphate (hence the name G-protein). This phosphate is nested in a cavity of their bodies. At some point in the demon's cycle, that cavity acquires catalytic power and splits the phosphate tetrahedra. The guanine trisphosphate here has the same function as the adenosine trisphospate (ATP) in the case of the demons we saw before; the splitting of the tetrahedra sets free a sizable package of energy, from which the demon draws his drop of negative entropy. There is this difference, though: the drop comes from within the demon—out of his sleeve, as it were.

And it does so with good timing. Timing is everything in sleight of hand, of course—all that phosphate energy wouldn't do much good if it weren't released at the right moment in the cognition cycle. So, we have every reason to believe that the molecular demon here has a built-in timer. Such a device need not be all that accurate; all it has to do is ensure that the energy is handy when the demon meets up with I_{m2}. And indeed, the timer starts ticking the moment the G-demon leaves the rhodopsin, then sparks off his phosphate-splitting machinery in good time for him to pay the thermodynamic price for his act of cognition.

In case you wondered about the timer, such devices are not in short supply in biomolecules. Indeed, they are not in short supply anywhere in nature. They are found at all levels of organization of the universe, from the enormous pulsating quasars in the sky down to the small radioactive atoms. The timepiece in the G-proteins is an integral part of the mechanism whereby these molecules cyclically change gestalt. The clock is set in motion here by the electromagnetic field effect that implements the information transmission to I_{m1}—the very electrostatic forces that give the push for that demon's shift in gestalt. His second gestalt has the cavity I mentioned before, in which the guanine trisphosphate is housed. And as soon as the demon adopts that gestalt, the phosphate occupies the cavity, because its concentration inside the cell is high.

The cavity also houses the catalytic machinery for phosphate splitting. So well before he gets to I_{m2}, virtually from the moment go, the demon has the wherewithal to pay the thermodynamic piper. At the end of
his wandering, he passes on the photon information, pays the piper in
phosphate currency, shifts gestalt—all in short order—and then, more
leisurely, returns to the rhodopsin for more of the same (figure 6.1).

The Ubiquitous Membrane Demon

In short, our little demon here is both information carrier and
negative-entropy provider—he brings his own thermodynamic ticket
to the cognition game. Besides, he boasts still another merit: he is in
thermodynamic harmony with the lipid environment of the cell membrane. So although made of a protein, he is entirely at home in that
fatty medium and glides in it laterally as if on greased rails.

No wonder he became one of Evolution's favorites. She used him
again and again throughout the ages for all sorts of cell communications, and long before there were brains, we find him in endocrine systems involving adrenaline, histamine, glucagon, antidiuretic hormone,
parathyroid hormone, luteinizing hormone, follicle-stimulating hormone, thyroid-stimulating hormone, angiotensin, prostaglandins,
photoreceptors, odor receptors, and neuronal synapses. I put the neuronal systems here at the end to keep things in perspective.

That the G-demon is so widely used speaks of a deep-seated unity
in Evolution's designs—something one would hardly have expected
from outside appearances. Indeed, despite a century and a half of neurophysiology and endocrine physiology, no one had a clue that these
fields had a key element in common. The workers in those fields had
merrily gone their separate ways, and there was no helping it. The
available tools of investigation didn't pierce much beneath the outward
forms. It was only with the coming of molecular biology that the various I_{m1} DNA code-scripts were read, and only in the 1980s and 1990s
did it become clear that brain and endocrine systems have a good deal
in common.

CHAPTER SEVEN

Molecular Sensing

A Direct Line from Nose to Cortex

We turn now to another sensory system in which the G-demon has a hand, our sense of smell. This is older than the visual sense. It has its feelers in the coarse molecular sphere rather than the quantum world, though it leaves nothing to be desired in sensitivity or chromatic nuance. Smell is a strongly evocative sense, with roots reaching deep into the lower and older parts of the brain; odors can call up long-forgotten memories and deep emotions. We can distinguish some 10,000 scents—a bouquet so rich that not even Proust had enough words for it.

Odors are airborne molecules. They dissolve in the nasal mucus and bind to specialized proteins in the olfactory sensory cells. These cells are stationed high up in the back of our nose (see figure 7.1). They are ensconced in the nasal epithelial lining, and their hairlike extensions act as antennas.

It is in these extensions that the odor-binding proteins are housed. But let me call them by their rightful name, "sensory demons," for like their peers in the eyes, these proteins are full-fledged cognitive entities. They recognize the odor molecules, picking them out of a throng of millions; each sensory-demon type recognizes a particular class of odor molecules. They are the heart of the sensing process and set in motion the sequence of events leading to the generation of electrical signals, the bits from which our brain composes its pictures of the world.

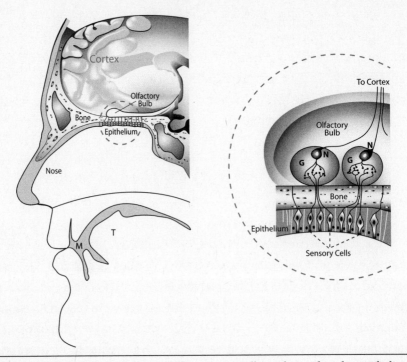

Figure 7.1. Our Olfactory System. The sensory cells are located in the epithelium of the nose cavity right under the olfactory bulb, the first information processing station of the brain. (*M*) mouth cavity; (*T*) tongue. *Inset:* A cutout view showing the sensory cells in the nasal epithelium and their axons traversing the skull bone and making synaptic contact with dendrites of neurons (*N*) of the olfactory bulb. The axons of these neurons go directly to the cortex. The neurons form anatomically distinct groupings, the "glomeruli" (*G*), in the olfactory bulb; only one out of many neurons in each glomerulus is shown.

And here is where the dependable G-demon comes in. At rest, he is fastened to a sensory demon. But that leisurely status quo ends when an odor molecule binds to that demon—the binding sets him free. So when a suitable odor molecule comes along, he takes his place among the members of the local I_M chain.* After that, things go on the old jog trot way (see figures 3.4 and 4.2), ending with the production of digital electrical signals that travel along the axon of the sensory cell to the centers of the brain.

---

*I_{m2}, the next member in the chain, is a membrane enzyme here, which produces the nucleotide cyclic AMP (I_{m3}) that opens the channels in the sensory cell.

The sensing process here is not as fast as in the visual cells—there are no superposed quantum waves seeking out the path for optimal information transfer in odor demons. But the electrical signals haven't far to go here; the first odor information-processing station of the brain is right above the nose. And from there, the electrical signals are sent straight on to the cerebral cortex (figure 7.1).

A Thousand Information Channels of Smell

But how does the brain keep the information of 10,000 scents apart? How many odor information lines are there?

The answer was found through a groundbreaking series of experiments by the molecular biologists Richard Axel and Linda Buck. They made a search for genes in the human DNA for odor-sensing proteins and found about a thousand of them. And there are about as many different sensory cells in the human nose, each cell expressing one odor-sensing protein type. Thus, given the fact that the axons of these cells go unbranched to the cortex, there would be enough discrete information lines for a thousand scents.

The brain areas concerned with smell exhibit a good deal of order in this regard, though we would hardly expect that from the helter-skelter situation in the nose, where the various sensory-cell types are deployed at random. However, on the other side of the skull bone, in the olfactory bulb, they sort themselves out, and the axons from the cell types expressing the same sensory protein converge onto the same neuronal cluster (glomerulus) (figure 7.1). That sort of simple topographic arrangement is repeated at the pyramidal-cell layer of the cortex—alike connects to alike.

Buck and colleagues more recently took the analysis a step further in mice. They genetically engineered mice that would express a marker protein in the olfactory sensory cells. The gene for the marker was inserted into embryonic stem cells next to the coding region of the gene for the sensory protein, and from those stem cells mice developed that expressed the marker concurrently with the sensory protein. The marker passed transynaptically from neuron to neuron, outlining the information channels from stem to stern. Two such channels are schematically

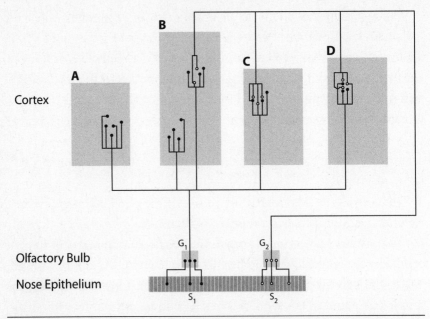

Figure 7.2. Topography of Smell Information in the Brain (Mouse). A diagram of the neuronal connections of two sensory cells (S_1, S_2), expressing distinct sensory demons (*black* and *white*). The two sensory cells are scattered among other sensory-cell types in the nose epithelium (only cells of two types are shown), and each type connects with neurons in the same glomerulus of the olfactory bulb (G_1, G_2). Those olfactory-bulb neurons, in turn, connect with cortical neurons that are clustered in four areas, the information-processing stations of the cortex: (*A*) anterior olfactory nucleus; (*B*), piriform cortex; (*C*), olfactory tubercule; (*D*), enthorhinal cortex. Only a few neurons in a cluster (which may contain several dozens) are shown.

represented in figure 7.2, illustrating how the corresponding domains at the sensory periphery map on the olfactory bulb and cortex.

Mapping, Coding, and Synonymity

The tracking down of the odor-information channels and the simple brain topography it uncovered was a major step on the way to our understanding of how the brain forges a world representation from the incoming information. But let's stop a moment here to ponder what that topography purports physiologically.

The first thing that ought to give us pause is the large number of genes for smell. The human DNA contains about 30,000 genes, and about 1,000 of them are for smell; to wit, 3 percent of our genome is devoted to smell! Nothing could speak more eloquently for the importance of this sense. But together with the brain topography, it speaks just as much for the simplicity of sensory coding: the peripheral domains of the thousand odor-information channels map in a straightforward manner on the cortex, and corresponding domains correlate forthrightly geometrically with one another.

Such simplicity sets the analyst's heart aflutter, holding out hope that we may decipher Evolution's sensory scheme. No topological tortuosities here, no mathematical complexities like those cryptographers concoct to keep military or banking information under wraps. Evolution just doesn't go for such convoluted tricks—or if I may paraphrase an Einsteinian mot, "the lady is subtle but she is not mean."

The coding here has much of the simplicity of that in a teletypewriter. That machine has gone out of fashion, though it's worth calling it back because it allows us to see the nitty-gritty of coding, the mapping of information domains, at a glance. In the teletypewriter system a message in English is transformed by one machine into a pattern of holes punched into a tape, then into a sequence of electrical pulses in a telegraph wire, and back into English by another machine. The patterns of holes in the two machines correlate geometrically with each other, and in each transformation the sequence of domains (symbols) is ordered according to a set of logical rules. That set of rules embodies the code.

Now if the code is stringent—and in a good teletypewriter it is—the message that is fed in will come out unadulterated at the other end of the line. If we make a list of all the symbols, we see that for each English symbol there is one pattern of holes and one pattern of electric pulses. It is a code with a mapping of one-to-one (see figure 7.3A).

That sort of mapping has no gray areas—the code here is so stringent that it leaves no room for ambiguity. But that's not a must. A sloppier code can still yield a satisfactory result in communication. In fact, we use such sloppy coding all the time in our daily verbal communications and get away with it; we use words with more than one meaning.

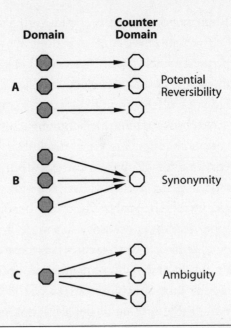

Figure 7.3. The Three Coding Modes. (*A*) one-to-one mapping; each element of the domain has a corresponding element on the counterdomain. This type of coding precludes synonymity and ambiguity and is potentially reversible. (*B*) convergent mapping; more than one element of the domain maps as one element of the counterdomain. This type of coding gives rise to synonymity. (*C*) divergent mapping; a coding that gives rise to ambiguities. The arrows indicate the direction of the information flow. (From W. R. Loewenstein 2000.)

We call this synonymity, but what it comes down to is a coding with a mapping ratio higher than one (figure 7.3B).

Molecular Sensory Synonymity

There is thus in practice nothing wrong with a bit of imperfect mapping (though professional cryptographers will look down on it and call it "degenerate coding"). Indeed, convergent mapping is the hallmark of a rich language. Without it our speech and writings would be dull and barren. English, with its roots in German, French, and Latin, abounds with it. And so does our sense of smell.

In the olfactory system convergent mapping starts right at the beginning of the information channel, at the interface between odor molecule and sensory demon. This brings us back to the nub of sensing, the cognition act of the sensory demon. Like all protein demons operating in the molecular world, this one recognizes a molecule by its shape. He carries the selection standards in his binding pocket and, depending on how well a molecule fits in there, it is either selected or rejected.

For a selection, the fit needs to be good but not perfect. There is really no such thing as a perfect fit between molecules—there is always an atom sticking out here and there, affording some elbow room. So, if you have a spate of look-alikes, the chances are that more than one will answer the purpose. And that is exactly what happens here.

Take, for instance, the alcohol and aldehyde molecules floating in the air. Many are odor molecules—they are of the sort a rose or a peach puts out. Some are fragrant as new-mown hay, others flowery or fruity, others a bit rank. When they come, they do so in droves—one kind or the other always seems to be around. And they all look as if they were tarred with the same brush: carbon chains of assorted length (see figure 7.4). So when

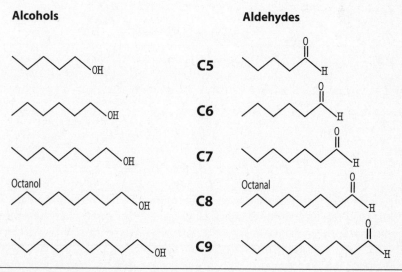

Figure 7.4. Odor Molecules. A sampling of alcohols and aldehydes, simple carbon chains with five to nine carbons.

they differ by only a piddling few angstroms, even the most able demon would be hard-pressed to tell the chains apart. And indeed, the odor-sensing demons don't. They typically respond to more than one of them.

A look at their binding pockets tells us why: the α-helices there contain quite a few hydrophobic amino acids that are not electrically charged. Such nonpolar entities allow for some playroom. For example, the demon for octanal, a rather smelly molecule (an aldehyde), offers up five such amino acids for interaction with the nonpolar body of the carbon chains (see figure 7.5).

So imagine an odor molecule entering the pocket. Buffeted by the thermals in there, it constantly bounces on and off the amino-acid lining, eventually coming to rest with its nonpolar carbon chain facing the five nonpolar amino acids. This makes for a somewhat shaky situation—it is merely the condition of lowest energy. The odor molecule is held in

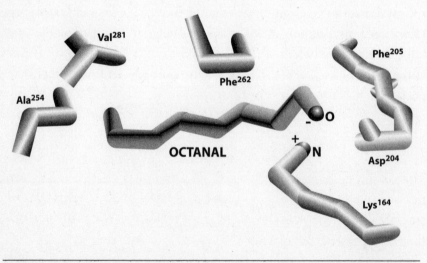

Figure 7.5. The Binding Pocket of the Octanal Demon. A computed model, showing the octanal molecule in its equilibrium position and the constraining amino-acid residues on the demon's α-helices. The negatively charged carbonyl oxygen at the tip of octanal (O) and the positively charged nitrogen (N) group on lysine[164] are at a distance of 2.83 angstroms, within range of strong electromagnetic attraction. The nonpolar body of the octanal carbon chain faces nonpolar amino-acid residues (five are shown), which are within range (about 4.3 angstroms) of hydrophobic interaction. (After M. S. Singer 2000.)

place by rather frail electrostatic forces in the pocket and hovers in there, rather than being firmly planted. The electrostatic forces individually sum up along the length of the molecule's carbon chain, but it takes at least a seven-carbon-long chain to make the situation halfway stable. Longer chains do better, but there is only so much room in the pocket.

Apart from those weak forces, the odor molecule is constrained by more focal electrostatic interactions. And here the underlying forces are strong, indeed, as strong as they come in the molecular world. These are interactions that take place between polar molecular portions, namely, the negatively charged oxygen head of the carbon chain—a carbonyl oxygen—and a positively charged nitrogen group on the lysine (Lys[164]) lining the pocket. Thus when the odor molecule comes to rest, its oxygen gets to within 2.8 angstroms of the lysine—close enough to be strongly attracted (figure 7.5).

This pinpoint electrostatic interaction narrows down the demon's choices. It holds the odor molecule by the head, as it were. The nonpolar rest of the odor molecule offers no opportunity for that sort of strong electrostatic interaction. That part of the molecule (which includes the tail end of the carbon chain) is constrained only by the overall limits of the demon's pocket. These are forgiving enough for some swing in the pocket.

Why Sensory Synonymity

Thanks to that swing, we have quite a few odor synonyms. Many are just variants of carbon-chain length or variants with a different group (say, a carbonyl or a bromo) dangling at the end. However, together they make for rich bouquets.*

---

*The existence of so many odor synonyms may come as a surprise—it is so out of proportion with our sensory experience. But that is conscious experience. A good part of the odor information coming into our brain is processed at the lower levels of the brain, and not all of it translates to conscious states (though it is biologically no less important). This situation is mirrored in our language. Our vocabulary of odors is notoriously small, and the number of our linguistic odor synonyms is totally out of proportion to that of the primary molecular synonyms.

But why are there so many synonyms? And how does the brain cope with the ambiguities that may result from them? The latter point will be foremost on the mind of the reader versed in information theory. But bear with me for another minute; I'll clear up the why question first, and the other will take care of itself.

To begin with, let's reformulate the question in Darwinian terms: What evolutionary pressures may have given rise to the synonymity here? That sort of tack already proved its worth when we considered the why of quantum sensing. If anything, the answer ought to come more easily now, as we have seen enough of Evolution's game to know that she is set in her ways. Those pressures are likely to be the same as the ones behind the synonymity in vision (chapter 5): the unrelenting pressures for an organism to gain information from the environment—the very same pressures that, long before there were brains around, had given rise to the basic systems of cellular communication.

The quest for information is as old as life, and the strategy for capturing it is as changeless as the Information Arrow. What changed was the information gain. This grew over the ages, at an ever-increasing rate, and there still is no plateau in sight—and there may well never be one. Much of the gain is owed to our senses. These untiringly feed the brain with information riding on quantum or molecular particles. That information stems from an inexhaustible reservoir, and Evolution cannily contrived to maximize the flow into the brain with minimal information outlay. In the case of the sense of smell, she lets us in on one of her tricks: she left some room for play in the odor demons' sensing pockets.

By this simple expedient, each demon responded to several species of molecules, rather than to only one. This effectively increases the information capacity of the brain; the total amount of environmental information available to the brain is greater than in the evolutionary alternative, a mapping at the level of the sensory demons of one to one. The beauty of it is that this increase in information capacity comes at no extra cost. Nor does it put undue burden on the developmental capacity of the brain, as it involves but the brain's outposts, parts that are not inside the cranial cavity where the premium on space is high.

Quantum Synonymity

At this point it is interesting to draw a parallel between the olfactory system and the visual one. Although in the latter the demons are very different sensors, there is synonymity, too, as diverse quantum particles, each with its own quantum signature, map onto one and the same sensory-demon domain.

The term *quantum synonymity* may at first sound a bit odd. But it is correct. It is just one more thing about the strange quantum world that flies in the face of our intuition. We are as unaware of that synonymity as we are of the particles' individuality; both are outside our immediate sensory experience, though now and then their individuality surfaces. The photon population in a ray of sun is a case in point. When these photons collide with the water molecules in the air, the various photon types segregate, as they are deflected differently, depending on their energy quantum. So what was once a drab particle hodgepodge now becomes a rainbow of glittering gems in the morning dew.

The various particle types—a multitude—all map on four rhodopsin domains (chapter 5), representing a synonymity even larger than that in the olfactory system. Evolution's method for getting synonymity here was no less elegant: she used a set of amino-acid dipoles in the rhodopsin molecule (figure 5.2a). That was a relatively costly strategy. Getting those dipoles deployed at exactly the right places costs more genetic information than getting some play room inside an odor demon's pocket. But overall, Evolution saved on the number of demon types. Whereas in olfaction she invested in a thousand types, in vision she got away with four.

The larger investment went with the territory, so to speak. In the molecular world, the particles just don't come in such smoothly graded quantal forms that can be neatly sorted out in fields of dipoles. The immensely larger particles in the molecular world need to be literally grasped, and that can only be done by relatively large molecules. So a thousand different protein molecules, although they constitute a thirtieth of all of Evolution's genetic investment, wasn't too high a price to pay for the huge gain in information input she got for the brain.

But she seems to have gotten even more out of that investment. Recent evidence indicates that the odor-sensing demons may double as molecules for cell-cell recognition. Their protein is transported along the sensory-cell axons and may provide a means for the neurons of the diverse odor-information channels to sort themselves out.

I don't know who audits her accounts, but whoever does, needs no red ink.

Harmless Double Entendres

Now to the second question: How does the brain cope with the dreaded ambiguity? I say dreaded because that's what communication engineers and cryptographers fear most in their information systems. The running engineering cliché is that synonymity leads to ambiguities in communication. Indeed, that is true for two-way communication. Turn around the information flow in a system with convergent mapping (figure 7.3B), and the relationship at the domains reverses: each domain now has several counterdomains—the mapping becomes divergent (figure 7.3C). In other words, each symbol here has more than one meaning.

This is why communication engineers won't touch that sort of mapping with a 10-foot pole. This is understandable because their communication systems—the teletypewriters, fax machines, telephones, Internet, etc.—are all designed for two-way information traffic. But Evolution doesn't work that way. Her communication systems, and certainly her sensory systems, are one-way. An odor demon or a visual one takes in information from the environment, but cannot put it out—his operation is thermodynamically irreversible.

In fact, all of the information lines of the brain are one-way streets, as the various stages, from the sensory stations to the retransmission stations along the way to the cortex, are all manned by irreversible demons. Thus, for Evolution, synonymity poses no problem—one-way convergent information streets are not ambiguous.

CHAPTER EIGHT

Electronic Transmission
of Biological Information

We now retrace our steps to take a second look at the electrons. So far we have touched on these quantum particles only in passing and only inasmuch as their energy fields are the substrate for our sense of vision. Electrons form the natural cortege of the atoms on the surface of the objects in our environment and are our sources of information about the world outside, the only ones our eyes have, though it's the incident photons, of course, not the electrons, that carry the information to the eyes. Inside our body, however, the electrons do carry information—and it would be surprising if they did not, as they are notoriously interactive with things molecular. They certainly are not the minimalists the neutrinos are. Indeed, they transmit vital information inside our cells, and in that endeavor they are the agents of one of the most basic biological information lines. The line developed early in evolution and comes as close to a solid-state electrical transmission line as anything ever does in the soggy biomatter. This makes it an interesting thing to study, but from our perspective here, just as interesting is the question of why that mode of transmission is not used in the information lines of the brain.

So, let's backtrack a little—*récouler pour mieux sortir.*

Evolution's Favorite Leptons

To begin with, consider the energy fields of the electrons of the surface atoms of a visual object. These fields are not carved in stone—nothing is in the blurry quantum world. But aside from that inherent blurriness, the electrons are freely mobile here and get exchanged between the atoms of the object. That exchange is what holds the atoms of a molecule together, and the forces involved determine the molecular structure—the distance between the atoms, their positioning, and their angles of orientation, the whole three-dimensional shooting match.

For decades those seminal forces had been the chemist's stamping ground and were dealt with empirically. But thanks to quantum theory, they can now be calculated and molecular structure predicted. That is one of the great successes of the theory, and it affords us a deep insight into the organization of matter—deeper than any number of X-rays ever could.

To be sure, we still need those wonderful revealing rays for large and complex molecules, like proteins. The quantum description of such molecules still escapes us; the forces involved require an enormous number of computations. But that's only a matter of time. With computers getting faster by leaps and bounds, even the largest and most involuted molecular structures will soon be explained the quantum way. Then we will also be able to fully describe those blurry electron scenes on environmental objects feasting our eyes—or to put it succinctly, to understand, not just see, what we see.

That and more, later. At this juncture we consider a more earthy matter: the electronic lines transmitting information in our body. Those lines are only a few hundred thousandths of a millimeter long, but they are of vital importance. They transmit the information our cells need to make use of the oxygen we breathe and to produce ATP—without those lines, our lungs would literally beat the air.

The lines go way back in evolution. They offer us a rare coup d'œil of how things once were on our planet when the differences between physics and biology were still small. Then the electrons in the emerging biomatter showed themselves plainly for what they are: *leptons.*

If this sounds like some of those little creatures of fairy tales, it's just a phonological accident. The name comes from the Greek, *leptos*: slight, slender. It was given to a class of lightweight entities in quantumland—the other members of that class are the neutrino, the muon, and the tauon. Think of them as tiny spinning tops with their axis of rotation pointing in the direction of their motion. The rotation can be either clockwise or counterclockwise—and so one speaks of "right"- or "left"-handed leptons.*

Leptons are fundamental particles, meaning that, as far as we know, they are not made of something smaller. The more massive protons and neutrons, which form the bulk of atomic nuclei, are composed of still smaller particles, the quarks—and those are fundamental in the sense the leptons are. Quarks resemble leptons in many ways, but they differ in this regard: they do not exist in free state; they are strongly bound within atomic nuclei. Leptons are bound, too, but by a much weaker force. In the case of the electrons, that force (the electromagnetic force) is several orders of magnitude weaker than the one binding the quarks (the nuclear force). The electrons are therefore readily pried loose—and being light and electrically charged (their charge is –1), they make swift information carriers.

In this world, that is as good as gold. Much of our technology stands on that lepton talent. Engineers take advantage of it to run electric motors; to send information through wires, transistors, and computer chips; to heat ovens and soldering and waffle irons; and to flood with light the dark of night.

Evolution, too, seized upon that lepton talent. She used it for *her* motors, the mitochondria—the tiny ATP-generating organelles that inhabit our cells. The information lines here are made of a special type of protein that is capable of donating or accepting electrons. And the pink of those proteins have iron or copper at their cores (see figure 8.1). They form chains where electrons flow down their gradients, much as they do in a metal wire.**

Electronic Information Transmission:
A Development Stumped

Those biological electronic lines transmit strong and clear signals— they are dependable short-distance communication lines. Yet their

---

*Particle spin, like electron charge, comes in fixed multiples of a unit. For historical reasons, the unit is ½ (the photon has spin 1). All leptons have spin ½.

**The information is carried by free electrons through the metal-protein parts of the chains and by hydrogen atoms or hydride ions through the soggier parts.

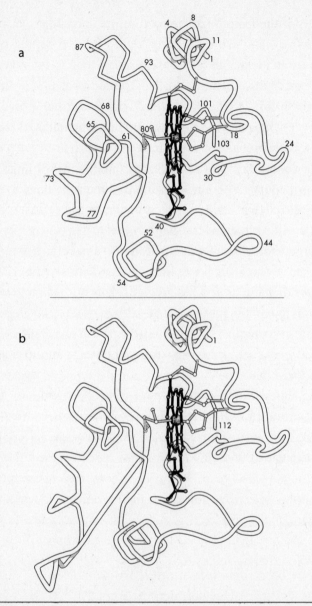

Figure 8.1. The Metal Protein Cytochrome C. The diagrams show the three-dimensional structure of the protein from (*a*) the tuna fish and (*b*) the bacterium *Rhodospirillum rubrum*. The iron (*Fe*) is at the center of a ring structure bonded to the polypeptide chain (at amino acids *18* and *80*). The iron oscillates between an electron-donating (ferrous) and electron-accepting (ferric) form, as it transmits electrons. Amino acid *1* is the amino terminal and *103* and *112* are the carboxyl terminals of the strings. (Drawn after X-ray crystallographic data from [a] R. Swanson et al. 1977 and [b] Salemme et al. 1973, 1977.)

development went only so far and would find but limited use over the aeons. Why?

It was certainly not for lack of speed. Those biological lines are as fast as hard-wired lines, if not faster. Nor was it for lack of information-transmission efficiency. Their metal proteins leave nothing to be desired in that respect. On the contrary, they are fit to be compared with tunnel diodes, as we shall see. More likely, the limitations of use were dictated by information economy. The genetic-information cost of lines made from metal protein isn't low, but is affordable when the lines are a few hundred thousandths of a millimeter long. But long-distance lines are another matter. The manufacturing cost rises steeply with length, and for lines long enough for intercellular communication, it becomes prohibitive. Thus, knowing Evolution's economic bent, the electronic mode may have never been a serious candidate in her laboratory for transmitting information between cells, let alone neurons.

Indeed, looking at today's metal proteins, we get the distinct impression that their development somehow got stumped. Take cytochrome c, for example, the kingpin of the electronic lines in mitochondria. This iron-containing protein came on the scene in single-cell times, already before plant and animal life went their separate ways. Yet it hasn't changed much. The cytochrome c of wheat doesn't look very different from our own. Both are single polypeptide strings about a hundred amino acids long, folded up similarly in three dimensions (see also the fish and bacterial cytochrome in figure 8.1). And they are functionally equivalent. Put them in the test tube, and the cytochrome c of wheat will substitute for our own without the electron flow and ATP production of the reconstituted protein chains missing a beat.

Two Old Batteries

The mitochondrial protein chains are the seat of an electromotive force—indeed, a respectable one amounting to a few tenths of a volt. Here the metal proteins are the functional—and, to some extent, the mechanistic—equivalents of the metal plates of an electrical battery.

Figure 8.2. Volta's Battery. A series of zinc (*Z*) and copper (or silver, *A*) plate pairs separated by pieces of pasteboard soaked in salt water (*black*). In the battery found in our cell organelles, the series is constituted by iron- and copper-protein pairs separated by salty cell water. (From Volta 1800.)

The parallels are most conspicuous with batteries of older vintage. Consider, for instance, the old-world model in figure 8.2. This is a battery with no frills: a series of zinc- and copper-plate pairs separated by pieces of pasteboard soaked in salt water. In the mitochondrial protein chains the elements are iron- and copper-protein pairs separated by salty cell water.

The model shown is, in fact, the oldest battery of human technology. It is Alessandro Volta's original model. I reproduced it exactly as he drew it at his presentation at the Royal Society of London 200 years ago. And I did so with no little pleasure, for it was this battery that opened the way to the discovery of the electron and to a deeper understanding of energy and matter.

I take the liberty of bringing in here a few bits of history that shed light on how that came to pass. In 1800 Volta, who was a professor at Pavia, Italy, presented his invention in a letter to Sir Joseph Banks, president of the Royal Society.* It was the right place and the right time; the far-reaching implications of the invention were soon recognized. Six years later the chemist Humphry Davy, in his Bakerian lecture at the same society, with Banks presiding, said, "It will be seen that Volta has presented us a key which promises to lay open the most mysterious recesses of nature." Davy then undertook his classical electrochemical experiments clarifying the nature of electricity. Ninety years later, the physicist Joseph Thompson (1897) took up the thread in experiments on the rays emitted by metal plates in the vacuum and discovered the electron.

---

*Volta's letter was written in French; he called his invention *pile*—a pile of metal plates—a term still used for battery in France (or *pila,* in Italy and Spanish-speaking countries).

CHAPTER NINE

The Random Generators of Biomolecular Complexity

We turn now to what we left dangling midway, the question of why Evolution didn't betake herself to the electronic transmission mode for neuronal communication. We learned that cytochrome c, the linchpin of that mode, stayed rather constant throughout the ages. Indeed, ever since the single-cell age, hardly a fold of it got out of place, which certainly couldn't be said of most other proteins. Those by and large have had a more eventful, at times turbulent, history. Typically, they undergo changeovers—adding a fold here, a pleat there, or a new curlicue—perennially metamorphosing from something old into something new.

The proteins involved in neuronal communications do so too, and if anything, at an even faster pace. So why won't the proteins involved in electronic communication do so? That question will take us to the DNA script, the accumulated repository of ancient biological information, for there lie the roots of protein progress. We will track down the causes of that progress and see that at bottom lies a quantum random generator of molecular form. And in regard to Lady Evolution's policy, we shall see that this is not unlike Mark Twain's: "I am all for progress. It's change I can't stand."

Genuine Transmogrification

I define first the meaning of *protein progress*. Progress is in the eye of the beholder, of course, but here, for once, we can come up with a criterion that is reasonably objective: the information intrinsic in protein structure, as given by Shannon's formula. That provides a measure directly in bits (see the appendix), and it may loosely serve also as a gauge of complexity.* One's first inclination may be to use a functional criterion here—to the evolutionist, that comes by reflex—but that invariably lands one in circularity.

Let's also settle here a matter of terminology. In dealing with protein progress, we need to distinguish between long-term changes in protein structure, the genetically determined changes related to progress, and short-term changes, the sort occurring in the course of the proteins' quotidian activities, their ever-recurring cognition cycles. For the latter sort, we will reserve the term *transfigurations* that we used before, and the others we will call *transmogrifications*. But I hasten to add that, unlike the transmogrifications of mythical lore, these are genuine and thermodynamically legal.

But how does one lay one's fingers on something that is so long drawn out in time? The protein transmogrifications occur on scales of hundred of thousands of years. So, for humans they are always things of the past. But fortunately they are not beyond recall, as we can turn to the DNA script. Used with circumspection, this old script serves as a historical record and is an endless source of hints and clues, especially when the scripts come from various phylogenetic rungs. Such scripts cover a lot of historical ground—they contain current and archaic information (exons and introns)—and the more distant the rungs are, the more ground the compounded scripts cover. This kind of investigation is a true-to-form Sherlockian pursuit and, in this day and age of automatic DNA sequencing machines, it involves only a modest amount of legwork. All one has to do is put corresponding DNA stretches from different organisms side by side, and the common denominator clues leap to the eye.

---

*The reader seasoned in information theory may prefer the measure of algorithmic information—also known as the Solomonoff-Kolmogoroff-Chaitin Complexity (see G. Chaitin 1987)—which, briefly put, is the number of bits in the most concise universal computer program generating a particular structure.

The first clues that did appear were the common family traits of proteins. There are several hundred thousand proteins now around (in our body alone, about a hundred thousand), and they turned out to share many genetic traits. This permitted the workers in that field to put some order in the protein crowd, a classification according to lineage, which by itself was an advance—a good reductionist move that eventually brought that unwieldy number of proteins down to about a thousand families.

But that's just a warm-up. Once the reductionist scalpel is out of its case, there is no holding it back. And it would soon cut deeper, opening up a panorama of dreamful beauty: an informational continuity in the biosphere stretching over billions of years—a panorama wherein all proteins, animal and plant, are members of the same family tree tracing back to a small aboriginal set, the first proteins on our planet.

So here we see our true-to-nature pedigree, and it would be hard to find one as blue-blooded anywhere in *Who's Who* or the *Almanach de Gotha*.

A Quantum Random Generator of Molecular Form

Let's wield that scalpel some more to dissect out the mechanisms of transmogrification. We needn't be too concerned where we cut, as from the coign of vantage the result above affords, we may choose any part of the protein family tree. And as for clues, there is no dearth of them in the DNA of that vast branchwork now overspreading the earth; Evolution left her fingerprints all over in the form of deletions, substitutions, and insertions. Such modifications in DNA script go under the collective name *mutations*. These are random events; they are caused by random hits of high-energy-photons or random hits of reactive molecules in the DNA environment.

We dwell on the photon-induced sort here. These mutations are owed to photons with enough punch to break a covalent bond—meaning ultraviolet, X-, or γ-rays (see figure 5.1). Such photons can change the otherwise stable information content of the DNA, and a single random hit may suffice for that. For example, an ultraviolet photon in the 260-nanometer-wavelength range will split a covalent bond of a DNA base,

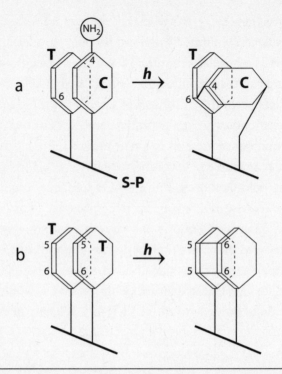

Figure 9.1. Mutations by Quantum Particles. Two examples of frequent DNA mutations owed to absorption of an ultraviolet photon: (*a*) fusion of two consecutive pyrimidine bases (through a 6–4 covalent linkage); (*b*) formation of a pyrimidine dimer (5–6 linkage). Such linked bases cannot be transcribed to RNA. *T*, thymine; *C*, cytosine; *S-P*, sugar-phosphate backbone; *h*, 260 nm photon.

causing that base to get fused to a neighboring one (see figure 9.1). Such a fused pair is not transcribed, so there will be a protein alteration.

That sort of mutation is but one of several an ultraviolet photon may produce—one of the most frequent; and the X- and γ-photons may produce their own, including breaks in the DNA sugar-phosphate backbone. Aside from such direct action, those photons can produce mutations in a roundabout way; they may alter molecules in the sap around the DNA and render them reactive. Both backbone and bases of the DNA are vulnerable to chemical attack, and it doesn't take much to warp the bases—just affixing, say, a methyl group onto the right base will do it.

Not that such mutations are everyday occurrences. The DNA is fairly well shielded against the ultraviolet by the water inside and outside cells, and X- and γ-photons, for their part, are normally rare. Furthermore, cells have elegant DNA-repair enzymes that offset DNA mutations. Those are powerful protein demons who will detect and excise a mutated DNA stretch and rejoin the strand. They will quickly cut out a little mutation, like the one in figure 9.1, and nip it in the bud.* They can cope even with the ultimate DNA damage, a double-stranded break of the DNA.

But good as those demons are over the long evolutionary haul, some DNA mutations will inevitably escape their action. And when they do, they have a good chance to leave a pockmark that goes all the way from DNA to RNA to protein. In fact, that chance is higher than we have the right to expect from statistical mechanics—much higher. Which poses a paradox if we take the traditional *DNA→RNA→Protein* paradigm at face value. However, the sober reality is different: rather than flowing in a straight line, the information flows in circles—and it does so twice, for good measure.

Such loops minimize errors in the information transmissions here—there is no way that the system could otherwise beat the statistical odds. Indeed, they are powerful error-suppressing feedback loops. We only have to look

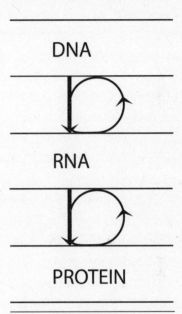

DNA

RNA

PROTEIN

Figure 9.2. Error-suppressing Feedback Loops in Information Transmission from DNA to Protein. Information flows back from the nascent RNA- and amino-acid chains via the corresponding polymerizing enzymes, minimizing errors in transcription and translation.

---

*A dramatic demonstration is provided by a human genetic disease, *xeroderma pigmentosum*, in which one of these enzymes is defective. The skin of the patients here abounds with ultraviolet-induced mutations of the sort in figure 9.1, and multiple skin cancers are the sad norm.

at the biochemical error rates to get a sense of that: 1:10,000–1: 1,000,000 for the stage DNA→RNA; and 1:10,000 for the stage RNA→Protein. These are the numbers of errors (E) in the information transmission at the two stages—that is, the number of mistakes made per number of DNA or RNA nucleotide units or amino-acid units (n) assembled at each stage—and they will readily translate to transmission probabilities $[p = (1-E)^n]$. Thus, for a protein of 1,000 amino acids, the probability of error-free transmission works out to be 0.91. This probability is so much higher than what is expected from the "$\sqrt{n}$" law of statistical mechanics that no number of exclamation marks would do it justice.

But just as stunning is the mutation apparatus as a whole, if we regard it from the quantum physics angle. Then we see what happens at the physics bottom, and the mutation apparatus shows itself for what it really is: a *quantum random generator of complex molecular form; to wit, a device that on one end takes up information from the quantum world and on the other churns out an ever-fresh variety of intricate macroscopic structures.*

There are other natural (autonomous) generators of complexity, of course, and no few in the nonliving world. Those produce well-varied molecular forms—clouds, snowflakes, crystals, rocks, water whirls, whirlwinds, and more. But all of them pale in complexity next to the proteins.

Advances in molecular biology in the past two decades have made available to us a fairly detailed picture of the inner workings of the random protein generator, including its evolutionary functions. However, there is nothing that would prepare us for the generator's time frame, which is on the scale of hundreds of thousands of years—and what could possibly come up to that in the human experience, where significant things happen in hours or days, lifetimes are measured in decades, genealogies in centuries, and all of recorded history in a few millennia?

Yet once we get used to thinking of a mechanism operating over so vast a time span, we begin to see this random generator as the hub of one of the oldest wheels of evolutionary progress. It has been running ever since the first proteins made their debut on Earth. And quantum potluck notwithstanding, it has been bringing forth new protein varieties with regularity now over some four billion years.

On such timescales there is an unmistakable rhythm to the generator. The rhythm is apparent on compounded DNA scripts, namely, on DNA stretches that do not encode proteins (such stretches are called introns) and on stretches that encode noncognitive protein (protein with a purely mechanical function). I leave out the stretches that encode proteins with cognitive or catalytic functions. Those are not useful for analysis here because they have lost much of the mutant information; the pressures of natural selection militate against the conservation of changes in amino-acid sequences that cause deficiencies of cognition. For a stretch of DNA of 1,200 base units—the average length of a gene—the rhythm works out to one protein variant every 200,000 years.*

So that, then, is the rate at which new gene information gets tried out in the laboratory of evolution—and the gene information for many brain proteins has been tried out the most.

We may press the analysis a little more to see how suitable that random generator is for the progress of a species. Take, for example, our own species, wherein such progress will depend on what happens to the DNA in the germ cells (the egg cells or the sperms). The arithmetic is easily worked out: at the mentioned rate, even with a modest population of 10,000 individuals, every possible DNA base-unit substitution will have been tried out in about a million years.

As the old saying goes, "God's mills grind slow but sure."

The Heuristics of Transmogrification

To us humans, those mills have always seemed things of the past. Even now, face to face with their physical embodiment, it isn't easy to rid ourselves of that misconception, because their impact in our lifetime is so slight. And not only has it been slight in our lifetime, but in the lifetime

---

*This estimate includes all mutations, the photon-induced and the molecule-induced ones. It is based on DNA stretches coding for fibrinopeptides (about 400 amino acids long), which have a mechanical function; they form the structure of the clots in our blood. The value is in satisfying agreement with estimates based on intron DNA. As for the reproducibility of the mutations, that is ensured by an information loop even tighter than the loops servicing the transmissions DNA→RNA and RNA→Protein mentioned before; the error rate DNA→DNA is 1:1,000,000,000.

of all humankind. We just haven't been around long enough. *Homo sapiens* came rather late onto the scene (figure 2.1), and by the time he did, most of his genes had already been formed.

Moreover, and more to the point, the information content of those genes had already been optimized through natural selection. In a broad evolutionary sense, this fine-tuning is the ultimate function of the random generator—its reason for being in the scheme of life.

This role isn't all that obvious. It is obscured, like the proverbial forest and the trees, by the countless bits and pieces in the DNA-RNA-Protein machinery. But the crisp focus of the information loupe brings it out. The products of the random generator then show themselves as information transforms of the mutant DNA, namely, three-dimensional transforms. And that three-dimensionality allows them to do what the original one-dimensional DNA information cannot: interact with the world of three dimensions, which is Evolution's laboratory.

In that laboratory the transforms can be tested, and they are, one by one, as they get turned out. Not many of them are up to Evolution's merciless standards, but the few that are get her ultimate accolade: their information lives on.

Such is the reward for optimal performance and, at the higher zoological echelons, it is given only if the winning transform happens to be in a sperm or egg cell—the subset of cells in our organism that carry our progeny. It is in that subset that the transmogrification really comes to fruition. What the random generator started, culminates there in a final and fateful act: the selection of the "good" information and the elimination of the "bad" information.

The Random Generator and Our Genetic Heritage

So the information content of a gene is constantly fine-tuned and, over the long run, optimized to fit the utilitarian biological context. Our own genes, except perhaps for a few that are being newly fledged, have all gone through such fine-tuning, and Evolution has had ample time to get rid of the rotten apples. Our DNA thus may not be as complete a historical record as one would wish for analytical purposes. Still, backed up with DNA scripts from lower evolutionary rungs, it makes a decent notebook—certainly good enough for chronicling what the random generator has wrought. However, we have to go to rungs sufficiently far down from us, as there is no point using the scripts from chimpanzees or other near neighbors, which resemble too much our own script (96 percent of the chimpanzee and human base-unit sequences of the hemoglobin gene, for example, are the same).

We are, after all, a species of ape that not so long ago climbed down from the trees. However, further down the evolutionary ladder, the gene information diverges more and the marks of the random generator are plainly seen. The mutation rates I gave before come from there. These are rates for individual genes. The estimates were taken from relatively short stretches of DNA, which was the standard way in the past. But in this day and age of large-scale science undertakings, like the sequencing of whole genomes, one can do better and determine the mutation rate of thousands of genes. Two such undertakings—the Human Genome Project and the Mouse Genome Project, aimed at determining the sequences of the roughly three billion base units in either genome—have recently been completed. They offer superb material for gauging the mutations in our genes.

The mouse has long been used as a genetics subject and as a model for human disease, and rightly so. Mice and men have an ancestor in common, a small mammal that split into two species toward the end of the dinosaur era—about 75 million years ago. This is just about right for analysis: long enough for mice and men to significantly diverge as species, but not too long for them to retain a significant amount of their common ancestral genes.

Indeed, our shared heritage shows itself clearly in the DNAs: long stretches of homologous base-unit sequences, 40 percent of the two DNA scripts, line up. As those homologous sequences are brought in register, the human-gene mutation rate falls out: an average of one mutated base unit out of every two.*

---

*The 40 percent of aligning DNAs comprise genes coding for protein as well as genes coding for RNAs that are involved in chromosome function; the 1:2 base-unit substitution-rate applies to both. The human genome contains 2.9 billion DNA base units and the mouse genome, 2.5 billion.

Thus, not much in our DNA has remained unhandled, and the little that has won't be so for long—the random generator never stops. Whatever this generator cranks out is incessantly tested in Evolution's laboratory and judged by its performance in the three-dimensional world in competition with other mutants. This is what makes her creatures so wonderfully adaptive. What for one set of conditions is the optimal performance, for another may not be. Let those conditions significantly change, and a new mutant moves up to the top.

An Algorithm Is No Substitute for a Demon

This concludes the story of the quantum random generator. But before I close the books, let me say a few words about its operational sequel, the selection of mutant information, and offer a note of caution.

The selection of mutants, the cornerstone of Darwinian evolution, has turned out to be an irresistible siren for mathematicians. The standard-free optimization here certainly invites mathematical modeling, as it looks offhand like the optimization problems mathematicians and engineers versed in system theory are familiar with, such as finding the fastest route between x number of cities; the optimal shape of an aircraft wing; the best way of packing cars into ship containers, wine glasses into boxes; and so on. For this sort of problem, expedient algorithmic solutions are on hand. These solutions were worked out with the help of an old and trusted friend of physicists: the "phase space."

Don't take the term *space* here too literally; it may actually be a bit misleading. It goes back to the 1890s. It doesn't denote a real space but an ideal (mathematical) one, an invention of Poincaré, which enables one to deal with both position and velocity in dynamic systems—to visualize the two simultaneously, as it were. This has an advantage over the ordinary coordinate system wherein only positions are seen. A point in multidimensional phase space represents the state of the system at a given moment—its entire instantaneous dynamic state. As the point moves with time, it traces a curve, the "phase-space trajectory," which defines not just the evolving position but the entire evolution of a dynamic system.

To specify that evolution—and that adds to the beauty of the Poincaré imagery—requires but a modest knowledge about the system. No need for complete records here; a limited set of the dynamic variables will do. The mutant DNA record is precisely such a set and, with today's speedy molecular-biological techniques, to obtain it is a simple matter—for all practical purposes, nature proffers the set on a silver platter. All of which understandably raises would-be modelers' hopes.

But I am afraid I must burst their bubble. Although their efforts are far-ranging and mathematically elegant, even to the point of launching algorithms for automatic evolutionary searches, there is a fundamental problem here that goes beyond pure mathematics: a problem rooted in the Second Law. It's one thing to specify the trajectory of the biological system, and another to actually *generate* it. And that's the asses' bridge. Specifying is something an algorithm might do, but generating in real space it cannot, without violating that highest of physics laws.

Let me explain. To generate something is an act of creation, and in the real world, a physics universe ruled by the Second Law, such an act cannot beg information (just as action cannot beg force). In the case of the real thing, there is no problem, and we have seen why: the requisite information is supplied by a macromolecular demon—either by an RNA or a protein demon, as the case may be. These Maxwellian entities have that information built into their bodies and, what is more, the wherewithal to restore it—and they do so over and over again, always punctiliously squaring the thermodynamics accounts (chapter 2).

No, the problem is with the algorithms. An algorithm may succeed in specifying an evolutionary fitness curve in phase space, perhaps even its complete topological structure, but regardless of how well its curve curves, it doesn't have what it takes to work *on its own* in the real world; it lacks the means to square the thermodynamics accounts. It cannot generate anything, let alone biomolecular complexity. One might as well ask it to pull the rabbit from the magician's hat.

You may ask why such algorithms work in engineering applications. Fair enough. The situation is quite different: the system to be optimized in those applications is not autonomous as a whole. There is invariably a human operator involved, and whether he uses a computer, a slide rule, an abacus, or a string of Mayan knots, his brain supplies the missing information, squaring the thermodynamics accounts in accordance with the cognition equation. Every single one of the many steps specified by those algorithms goes through the grind of the cognition equation; there is no escaping that, and the brain and its hordes of molecular demons have information enough to pay the thermodynamics piper.

In short, the arena of evolution is not a place for run-of-the-mill systems analysis. Those who turned to dynamic-system algorithms to solve the riddle of natural selection let themselves in for an enticing but illegitimate affair . . . and, well, after bingeing for so long on mathematics claret, it's time to sober up.

The Second Generator of Biomolecular Form

We come now to another biomolecular generator. This one also operates at the DNA level—and side by side with the quantum random generator. But while the quantum random generator propels biological evolution at a stately pace, this one does so by fits and starts.

However, there is method to this madness, and there is no mistaking it if you look at what comes out: intricately convoluted and involuted molecules—a degree of molecular complexity unmatched anywhere on Earth.

The first thing to be noted about this generator is that it ranges far. While the quantum random generator acts on individual DNA base units, this one does so on whole sets of them, even whole genes. Indeed, its action goes far beyond the mere substitution or deletion of DNA modules. It produces large-scale rearrangements in the DNA, altering the code script in profound ways.

I'll describe some common rearrangements in ascending order of their complexity. In one kind, DNA pieces get dislocated on the DNA helix and then linked together, giving rise to novel genes (see figure 9.3a). The new-fangled genes then are further worked on by the quantum random generator and fine-tuned—that old warhorse is always standing by.

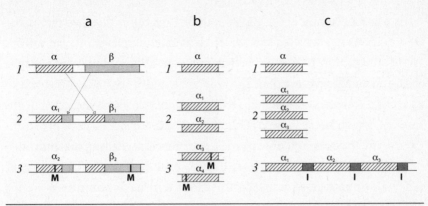

Figure 9.3. How New Genes Are Born from Old. Three examples of the second generator's operation. (*a*) shuffling of exons: *1*, the original genes, α and β; *2*, after the shuffling of two gene pieces; *3*, after point mutation, *M* (by the first generator). (*b*) gene duplication: *1*, original gene; *2*, replicas, α_1, α_2; *3*, after mutation. (*c*) gene composites from duplicates: *1*, original gene; *2*, replicated genes, and *3*, after their amalgamation. The replicas are flanked by introns, *I* (which are spliced out from the primary DNA transcript and provide a continuous reading frame in the replicas' protein translation).

Simple as it may seem, such reshuffling takes a good deal of information. These are coding pieces (exons), which are excised at exactly the right places on the DNA and then covalently bonded at the right ends. In addition, base sequences serving as "on" and "off" signals for translation are included at the ends. This is thus a well-concerted operation—something that, we know, calls for sophisticated demons. Indeed, the system here swarms with them.

In a variation on the foregoing theme, the new gene is duplicated a number of times (figure 9.3b). Each copy then is separately worked upon by the random generator and individually fine-tuned. This is how families of genes are generated. The members of such families, and we have already seen quite a few, are much alike. But because they have been individually tuned, that is, according to different optimization criteria, their products may have very different functions in the three-dimensional world. The four types of rhodopsins in the visual system, which have different wavelength optima, are cases in point.

In another variation, the gene duplicates are linked together, giving rise to a new, long gene (figure 9.3c). Such gene composites involve introns (noncoding DNA pieces) as well as exons and, if anything, they require even more information than the rearrangements mentioned previously. Not only is there coordinated DNA excision and juxtaposition, but the reading frame must be kept from being interrupted if the DNA composites are to act as functional genes. Here is where the introns come in.

These are DNA components that go back well over a billion years. They are interspersed between the exons, but have no protein-coding function. In fact, they are purged from the DNA transcript (the primary RNA) before this ever gets to the protein-translation stage. This is why molecular biologists originally called them "junk." But as it turned out, introns have their uses after all and play a crucial role in the transition from gene to protein: they constitute symmetrical modules in the gene's reading frame (jointly with the exon copies in the gene composite)— each exon copy is flanked by two introns of identical phase, allowing the composites to be translated continuously (figure 9.3c).

The second generator and its gene products play important physiological and evolutionary roles—the sheer magnitude of the information in those products speaks volumes. They make up a good part of our genome, and the larger ones are responsible for many of our higher organismic functions, including those of the brain.

Take, for example, the long gene composites. These encode a wide variety of membrane proteins, including some of the sensory demons at the brain periphery, as well as a number of demons of cell-cell recognition, which enable the diverse cell types in our organs to tell each other apart (see figure 9.4). Those gene composites supply the information for the higher cell organizations, embryonic and adult. They are involved in everything multicellular, from the formation of our skin, glands, kidneys, liver, heart, and so on, to the formation of our brain. Especially the brain, for there they encode not only cell-cell recognition proteins but also proteins controlling axon length and growth. They play an equally prominent role in our immune system, in the encoding of antibodies and complement factors. The modular architecture of these proteins clearly reflects the multiple duplications of their encoding DNAs; if we

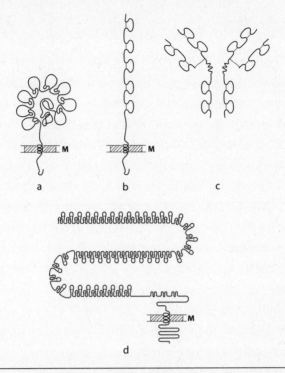

Figure 9.4. Modular Proteins Encoded by Duplicated Composite Genes. (*a*) sensory receptor for mannose; (*b*) neuronal cell-cell recognition demon, "N-Cam"; (*c*) antibody, "Immuno-globulin" (which shares with N-Cam a common gene motif); (*d*) neuronal cell-cell recognition demon, "Notch." *M*, cell membrane.

put these DNAs side by side with the DNAs encoding cell-cell cognition proteins of the brain, we see that the two protein types share a common gene motif (figure 9.4).

It is difficult to say precisely when the second generator started—its operation is spread out so widely in time. It may well have done so already in single-cell times. But it seems to have come into its own only after the transition from single-cell to multicelled organism, the transition named Grand Climacteric in figure 2.1; it was then that those gene composites matured and flourished. These long gene entities, in a sense, are the silent movers of the one-and-a-half-billion-year-old march of multicell evolution. And I don't think I go too far in saying that they are the ones who catapulted evolution over the multicell organismic threshold.

Complexity as a Windfall

The two random generators work hand in hand, and it is by dint of their close meshing that they bring forth that fabulous diversity and complexity of biomolecular form. But just as fabulous is their parsimony. We already got a sample of it at the input side of the first generator. Recall that this generator runs on quantum information coming in from the cosmos—information on the house, so to speak, whereas the second generator builds on that and arranges and rearranges the information products in all sorts of combinations. Among the two, they squeeze the information in the DNA to the last drop.

But the parsimony doesn't end there. If it did, there would not be a prayer for life. In fact, the parsimony is carried on to the generators' very output side, the one facing the three-dimensional world, where their products are sculpted in three dimensions. Here evolution once again shows her savvy. She gets the sculpting done for free: the three-dimensional shape of the products is their minimum-energy configuration—a sort of default configuration in computer terms.

She plays a combinatorial game here with linear information pieces, that is, DNA nucleotide strings that translate to an amino-acid string that folds up in three dimensions. Before the folding, the game is information-heavy and thermodynamically costly and follows the well-worn DNA→RNA→Protein groove. But the folding itself needs no further information input (nor supply of negative entropy), because the amino-acid string automatically folds into its three-dimensional ground-state configuration as it comes off the assembly line. The protein's shape, with all its crimps, crinkles, and undulations, is a windfall.

Ikats

The two random generators—their information structure, their information parsimony, even their dynamics—are so out of the ordinary that an analogy would be helpful. Alas, in all of human technology there is nothing that works that way—certainly not in three dimensions. However, if you are willing to make do with one less dimension, I can offer an analogy of sorts: the Ikat.

This is an old weaving art that once flourished along the silk trade route (China, Bokhara, Samarkand). Sadly it is disappearing, but you can still see it practiced here and there—Indonesia is a good place. And it is a feast for the eyes: from one silk fiber, the weaver makes an exquisite fabric; he feeds into the loom a single prestained thread of silk and, abracadabra, out of the warp and woof emerges a cloth with resplendent color play and complex design.

Figure 9.5. An Ikat. Below, a weaver and his loom. Inset (*a*) shows small segments of thread where red-stained portions alternate with unstained (waxed-over) portions.

But there is no bunkum here, of course. The secret plain and simple is information—programmed information. The program is in the thread. Before the weaving, the weaver or somebody else stained the thread so that its segments varied in color and length, according to a preset pattern—a cliché. That cliché holds the instructions for the cloth's design. The design comes out automatically in the weft, as the instructions are coded into the segment length and color sequence of the thread. And it is a straightforward coding. Indeed, it couldn't be more straightforward; it has a one-to-one mapping, leaving no room for ambiguities (see figure 7.3a). The weaver could be blind and the design would still come out the way it does—such is the power of good coding.

Yet this system allows for diversity. The cliché is easily varied: just a minor change—a touch of a different color, a slight shortening of a colored segment—and out comes a new design.

In the case of the three-dimensional protein fabrics, the oldest weaver tended the loom, making the most of the available information: a trifling substitution of a DNA base becomes a new protein involution; a minor repositioning of an exon, a new protein convolution; and a recombination of bases, a sea change.

CHAPTER TEN

The Ascent of the Digital Demons

After dealing with the two random generators of biomolecular progress, we are in a better position to tackle the question, why Evolution didn't use the electronic mode in her neuronal communication lines. We can now reformulate the question in terms of the second generator and ask why this ever-wheeling wheel of change would pass the metal proteins by. It would be presumptuous to try to guess Evolution's every whim, but in this particular instance we have a betting chance because she allows us to see a very long stretch of her history.

Quantum Electron Tunneling

The question of why the second generator passed up the metal proteins is all the more intriguing in view of the information efficiency of these biomolecules. We likened them to metal wires and metal plates, inasmuch as they transfer electrons. But they are actually better in that regard. Inside their solid-state atomic matrix, the speed and efficiency of electron transfer reach the extreme; they are comparable to the wonders of technology, the tunnel diodes.

I say wonders because the electrons in those high-tech devices bring off the seemingly impossible: they go over an energy barrier that is higher than the energy they have themselves. It's a bit like throwing balls through a wall instead of over it, and the balls go through as if the wall were so much thin air. Even the jargon used by engineers reflects that picture; the electrons are said to "tunnel through."

That is a simplism, of course. There is no solid wall if you scale things up from the minuscule dimensions of the quantum world, nor are the electrons mere balls. The wall is an electromagnetic barrier, and the electrons are quantum particles that can behave like waves. If you set them against such a barrier, those waves have a finite probability of appearing on the far side of it.

Strange as it may seem, that is the reality in the quantum world. And it is beyond doubt; a solid body of experimental physics and some of the great successes of high-tech industry attest to it. You need look no further than your desk computer. If it is fast, the chances are it contains devices in which the electrons are tunneling. There is really nothing strange about that behavior. It is the electrons' trait behavior; they merely act out here their innate lepton talent and heed the laws governing the quantum world.

And this is exactly what the electrons in the metal proteins do. This is relatively recent knowledge, and it came to light against all odds. It had long been known from the work of biochemists that the electron transfer in such proteins is highly temperature-dependent. So it was only natural to assume that the electron transfer was but a run-of-the-mill chemical process, and biologists were of no mind to hear, much less embrace, the idea of a quantum-wave phenomenon here. That is how matters stood for decades until John Hopfield, in a trenchant biophysical analysis, showed that the temperature dependence pertained to only a piece of the protein's action. Another piece, the most exciting one, had been missed: an action where the electrons tunnel through the protein.

Not that they do so through the entire protein structure. They tunnel only in a small ambit of the protein, a domain of 0.5–1 nanometer (in cytochrome c the domain is located at the almost exposed edge of the protein's metallic atom group; see figure 8.1). But this is where it's at. This is the solid-state atomic matrix of an otherwise flabby molecule, the locus where Evolution brought the speed of information transmission to the limit matter will bear.

The Electronic Cul-de-Sac

It must have taken a good deal of honing by the first generator to make that solid matrix fit into a protein molecule. But whatever honing happened was done and finished during the single-cell era. The DNA script tells us that in the following one and a half billion years the protein did not change significantly. Nor did the electronic mode of information transmission find new uses in the expanding biomass—a sad fate for a communication mode that was once the avant garde.

The reason probably lies in the conjugated structure of those proteins, a structure in which a metallic group of atoms is tightly knitted to the polypeptide chain. That group fits so closely into the protein's central crevice (figure 8.1) that there may have been no leeway for further evolutionary progress. In any event, those metallic atom groups are essentially planar and work their quantum wonders only over nanometer distances. It is difficult to see how they could possibly expand that range to the distances multicellular development would demand. The intercellular communications developing in the multicell era would call for a 10,000-fold increase in the range of information transmission—from the 1-nanometer magnitude to the 10-micrometer one—and that only for starters. To think of engineering a leap like that with the metal proteins is crying for the moon.

The Rise of the Digital Demons

The leap eventually came to pass, of course, but it was brought about by a breed of protein demons very different from the metal proteins. Some specimens of this breed we have already met. They are large proteins with a distinctive physique: their subunits are concentrically arrayed about an axis, forming a water channel (figures 3.2 and 3.3), and in the cases we are concerned with here the channel lets small inorganic ions through when it opens up, producing a digital electrical signal.

Macromolecular demons are never without the quirkish, but these have more than the usual dose. Like all members of that fraternity, they are cognitive entities. But they are doubly so: they discern individual ions and the voltage at their membrane locale. Both cognitions are attained at nominal information cost (perhaps no more than the irreducible thermodynamics cost stipulated by the cognition equation). Indeed, everything smells of economy here. Even the passage of the ions through the open channel comes at no cost to the demons; the driving force is paid for by the metabolism of the cells in which the demons reside—it's on the house, as it were.

The discerning of membrane voltage is the task of a voltage sensor that is built into the demons' protein structure. That sensor controls the open state of the channel, and when the local membrane voltage reaches a certain level, the channel opens up. And it does so at full throttle—it either opens fully or not at all. That all-or-nothing behavior is the hallmark of these demons. It is what gives their electrical signal its digital character, and hereafter we will call them *digital demons* for short.

These demons are relative newcomers on the evolutionary scene. They made their debut only around the Grand Climacteric (figure 2.1), but they would rapidly move up in the world. It is easy to see why if we watch them in action in the surface membrane of a long neuronal dendrite or axon. There, they occur at high density, so when one demon is activated and opens his channel, the chances are that a number of others will be within voltage earshot and open in turn. So one channel opening follows another until the digital electrical signal has swept over the entire length of the dendrite or axon.

The information elegance and self-sufficiency of this performance would secure the demons' evolutionary ascendancy. Today, we find these demons everywhere in the animal kingdom where fast long-distance cellular communication and fast computations take place.

Do Plants Have Digital Demons, Too?

In the light of what we discussed in chapter 3, we may assume that the digital demons arose under the evolutionary pressures on multicellular

Figure 10.1. Venus Flytrap and Prey.

organisms for speed of reaction. In the plant kingdom such pressures are generally lacking, and so are the digital demons. But there are interesting exceptions: certain plants growing in environments where nitrogen and phosphorus are scant. At some time these plants must have come under the mentioned pressures, as they adapted by feeding on insects to get the missing elements. A case in point is the Venus flytrap, a plant growing in the sandy bogs of North and South Carolina. This insect-eating plant has a set of special leaves forming a clamlike structure that snaps shut upon contact with prey (figure 10.1). The cells in these leaves generate electrical signals when the prey touches the hairs of the leaves.

Insect-eating plants were a lifelong interest of Darwin. I had the opportunity recently to read a letter he wrote to his friend Joseph Hooker on this subject (ca. 1866). He wrote that he was restocking his greenhouse and embarking on a study of "curious and experimental plants," touch-sensitive mimosas, insect-eating sundews, and pitcher plants, and elatedly added, "I can buy pitcher plants for only 10s.6!"

Well, these days we have to pay a little more, but a variety of such plants is now available from specialty greenhouses. They are easily potted and will regale you with quite a spectacle. In the case of a Venus flytrap, you won't have to wait long: a victim descends, lands, crawls to the set of special leaves . . . and with one fell swoop, it is trapped.

The leaf hairs act as mechano-electric transducers. They generate an electrical current in the leaf locale, and when the current reaches a threshold value, it sets off a larger electrical signal that propagates over the leaf. The mechanisms underlying these signals are not yet understood, nor is how they tie in with the mechanics that snap the leaves shut. Probing the cells with microelectrodes here is harder than in animal cells—the cellulose cover keeps the cells' signal bit tightly under wraps.

It would be of great interest to find out whether these plants have developed something akin to sensory or digital demons. All we have at this time is the hint that the mentioned electrical signals depend on calcium ions.

CHAPTER ELEVEN

The Second Information Arrow and Its Astonishing Dénouement: Consciousness

Let us briefly review the characters who have come on stage so far. It's a motley troupe, but they have this in common: they all are accomplished Maxwellian molecular demons and all are carriers of the Information Arrow. The first to appear were the chlorophylls and carotenes. This was the vanguard that let the arrow in from the quantum world. Next came the demons of more heft, the cellular proteins, who lifted the arrow above the molecular noise inside cells. Then came the intercellular-communication demons, who dispersed it throughout the multicellular mass. Eventually, and partly overlapping with the former, came the sensory and digital demons. These raised the arrow to new heights, setting in motion a seemingly unstoppable neuronal development with a dramatic dénouement: consciousness.

Like all things biological, consciousness probably developed gradually, and we take it for granted that it culminated with us humans—though we are hardly an impartial crowd. However, this much we may take for granted: it is limited to organisms with neurons. Few, except the most died-in-the-wool pantheist, would hold that single-cell organisms have consciousness, or that plants have (and I include the insect-eating ones). This implies that it began after the Grand Climacteric, because neurons made their debut only then (figure 2.1). But we wallow in ignorance about when in that relatively short, one-and-a half-billion-year stretch of evolution after the Grand Climacteric consciousness first appeared; neurophysiology, and all the neurosciences combined, can tell us nothing in that regard.

However, there is an asset in our arsenal we still haven't used: the information loupe. And it holds out hope. In its sharp focus, an auspicious place in that stretch is revealed a few hundred million years past the Grand Climacteric mark. It is a place, we shall see, where Evolution changed gears in her neuronal enterprise and started a new strategy, taking advantage of a peculiarity in the structure of time of the molecular world.

The Structure of Time

So once again the physics of time creeps into our plot. It already did so earlier in connection with consciousness, in our awareness of the passing of time. Recall that what we perceive as the flow of time is owed to the statistical interplay between the myriads of components of macroscopic systems, the immense number of molecules moving pell-mell there. When left on their own, such systems are immensely more likely to get disordered than ordered, and it is this relentless march to disorder that we perceive as the passing of time (chapter 1).

That perception comes down to a series of instants of increasing disorder. Think of them as a series of snapshots strung out on a line—it's what usually gets plotted on an x, y diagram from left to right. But such neat lines can be deceptive. They hide an interesting, and from the evolutionary point of view the most interesting, aspect of time: a structure. Depending on the direction we look, the line has a different texture—a sort of grain. In the forward direction, that is, from left to right—or what we usually call the future—the events in the snapshots are undetermined, whereas backwards, the past, they are determined.

We may normally not be conscious of this structural difference. Nevertheless, it influences, indeed directs, our behavior in a thousand and one ways—our short- and long-range planning; our rainy-day policies; our betting on horses, the stock market, business ventures, and so on. In all of that the grain in time structure is taken for granted, as we hold it implicit that we can influence the run of things in the forward direction, but not in the opposite one.

So those popular x, y diagrams adorning our classrooms and corporate boardrooms are really not so smooth and simple. The events they chart have an inherent structure—an inherent rightness and leftness.

That has been the way of things in the macroscopic domain of our universe from the moment go. The universe was born in an instant of high information (chapter 1)—an exceptional moment. In fact, so exceptional was it that even now, 13.7 billion years later, the effects are still working themselves through the system.

The Evolutionary Niche in the
Structure of Time: A Hypothesis

So, regarding the macroscopic domain, especially the molecular one, there are two lessons to take home: the structure of time was there ever since the first instants of the universe, and of more immediate interest, it was there when Evolution began her neuron enterprise.

Let me now put my cards on the table. I propose that *the time structure in the molecular domain was the evolutionary niche for neuron development, including its dénouement, and that the very skewness of the structure spurred on that development.*

I envision a development with two major stages. In the first, starting near the Grand Climacteric, neurons entered into small-scale associations. The members of those associations (initially but two or three) formed information loops—positively and negatively controlled cybernetic loops. In information terms, this development was still rather low-key. It brought some information gains, but they were relatively modest, just enough for fast reflex reactions. The descendants of those primordial loops are still in us today, forming the basis of our spinal reflexes or their equivalents in the cranial nerves.

The second stage came several hundred million years later, perhaps a fifth down the post-climacteric traverse. It was by far more ambitious. Vast numbers of neurons were then on hand, providing the wherewithal for an association on a much larger scale, the *neuron trellis*. This association had much more information power than the old cybernetic loops. It had a substantial memory and computer capacity, enough to take advantage of recurrent information in the rear of the time structure, to render the undetermined fore more determined—or put in less abstract terms, it could calculate, from recurrent information of the past, the odds of future occurrences.

In short, the neuron trellis is an anticipation machine.

Forecognition

Lest the image of Sybil intrude here, I'll spell out the machine's action more completely. It does three things: (1) it stores past information, (2) it retrieves recurrent pieces from its memory store, and (3) it computes from their frequency the probability of future occurrences. All three are perfectly legitimate actions under the laws of physics. They are on the whole not very different from what a physicist does in statistical mathematics when she computes the probability amplitude of a system's states. Thermodynamically, they are entirely aboveboard. The information gains they cause, though huge, are balanced out according to the cognition equation (chapter 2), and the trellis system pays the thermodynamic tithe.

Thus this kind of anticipating the future has nothing to do with the talent of Sybil or her like. But to avoid any possible confusion, we'll steer clear of words like *foreseeing* and coin the term *forecognition*.* This term also is physiologically more apt, as it covers all our senses, not only vision; but above all, it keeps us mindful of the fact that the cognition equation rules the roost everywhere in our skewed time structure.

The thermodynamic cost of forecognition is immense. Even in the case of modest forecognitions, we are dealing with negative entropies many orders of magnitude higher than in the case of the reflex actions of neuronal loops. So it was probably not before well into the multicell era that there was enough information capital on hand to defray the

---

*As all the English-vocabulary slots here had been taken for things outside the pale of science, a new word was needed.

cost. But once that hurdle was overcome and the first trellises were launched, there was no stopping them. Under the perennial evolutionary pressure for fast reactions, ever larger and better trellises would develop, giving their owners a decisive edge in the struggle for life.

And what could have been more decisive than forecognition in that struggle! These creatures were able to jump before it got too hot, run away before a predator pounced, gather food before hunger set in, or tell from the wag of a tail or from a mien whether it was friend or foe.

And so it went for about a billion years, the trellis sprangling, its memory and computer capacity expanding until nature's darling appeared. With a big bulge in front of his head, he must have been a strange sight for the other hominids. But the bulge made the skull cavity large enough to accommodate a trellis of a trillion neurons. And with that he was able to reach into the future as no one ever had before: he could compute at a glance at what angle an arrow would hit the mark, or tell from a gathering cloud that the weather would change, or tell from the position of stars when the winter would come and how many kernels of corn to plant to survive it . . . and one fine morning, he would get the knack of $t + dt$ and reach for the sky.

CHAPTER TWELVE

How to Represent the World

The Universal Turing Machine

In 1936, a young mathematics student at King's College, Cambridge, formulated the theoretical basis for a machine that could perform all sorts of mathematical tasks. That formulation would cast a wide net: it would bespread mathematics and physics and biology—even such a seemingly way-off field as brain physiology. The student was Alan Turing, the very same who, three years later at Bletchley Park, the British wartime cryptography headquarters, would crack the "Enigma" code of Hitler's armies. But for the time being, at Cambridge, he was engaged in a more laid-back pursuit: whether and how mathematical assertions can be proven. What he was after was the inner essence of the mathematical process, and he came to the conclusion that anything that has a mathematical solution, anything computable at all, could be computed by a simple machine equipped with a one-dimensional tape bearing a binary code. That machine has come to be known as the Universal Turing Machine.

It was an old dream come true: a machine that could perform complex intellectual tasks. That dream goes back to Pascal and Leibniz. But not until Turing was there any machine even remotely up to performing complex computations—the only thing capable of running an algorithm for that was a mathematician's brain.

But that soon would change: the first machines of this sort were built within nine years of Turing's seminal work. Those machines used the just-then-invented electronic vacuum tubes and were rather bulky and cumbersome. But nimbler ones were not far behind, and chances are there is one right now on your desk. One of the bigger and fast ones I had the opportunity to see in operation occupied 76 large cabinets at the Sandia National Laboratory, New Mexico—it's a giant version of your desktop computer (and made with the same Intel Pentium Pro chips). In the blink of an eye, it could do what would take an army of mathematicians thousands of years: a trillion operations per second.* And it would be just as reliable, or more so.

Well, it better be; those who wrote the International Nuclear Test-ban Treaty were banking on it! But it is not the speed or reliability of Turing's machine that concerns us here, but rather the insight it gives us into the process of computation—and all the more when that process, as Turing daringly set forth, takes place inside our brain.

Let's take a look at Turing's original conception of the machine. Rather than electronic parts, it is made of simple mechanical ones that are in plain view: a long, possibly infinite, paper tape and a mechanical "head" (figure 12.1). The tape is divided into little squares with one of two symbols written on them, say, a 1 or a 0. The head examines one square at a time, reads the symbol, and then either remains in the same state or changes to a new one, depending on what it reads. It moves the tape backward and forward, according to simple, unambiguous rules.

At bottom, and from a purely mathematical view, the machine is only a device for transforming one string of symbols into another, according to a set of rules. And this is how Turing originally saw it. The physical nature of the states of the head is unimportant. In Turing's original conception the states were mechanical. But they could just as well be electrical or quantal—we'll come back to this point further on when we deal with real heads. What matters mathematically is that the states be finite in number and follow predetermined rules. What the head does to the tape depends on the state the head is in when a particular frame gets to the head, and what it will do next depends on what it

---

*An upgraded version has just been built in that laboratory, which is two orders of magnitude faster.

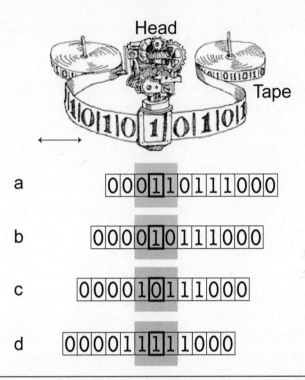

Figure 12.1. A Universal Turing Machine. It bears a set of instructions on a tape before the numerical data to be computed start. The window of the head displays the binary number labeling the present state of the head.

has just done, its past history. This way the symbols on the tape can serve as input to the machine and play a role in determining its subsequent action—the machine has a memory, and the rules are recursive.

Thus, the head can do one of three things: nothing, change a 1 to a 0, or change a 0 to a 1; and it can move one square to the right or to the left or not at all. And so it goes, the head reading, writing, changing marks, and bobbing back and forth along the tape, until it reaches a point where the square it is currently reading, and its symbol, stay the same. Then it stops. The machine has done its headwork, and the result is displayed on the output side of the tape.

Such a machine will perform any and all mathematical operations; no need for a separate machine to add, multiply, divide, extract square roots, and so on. Give it an appropriate program, and it will work out any computable mathematical function—hence the name Universal. And in this sense, all general-purpose computers today are Universal Turing Machines. They may have all sorts of paraphernalia—TV screens, graph plotters, editing devices, printers, memories that are not just one-dimensional tapes—but their heart and soul is Turing's handmaid.

Rendering the World by Computer

With a little more ado, Turing's Universal Machine will even go one better: it will compute objects of the outside world—or more to the point, it will render them in the form our senses perceive them. In other words, it will mimic the outside world.

That is the physics flip side of Turing's mathematical brainchild, though it was something physicists were slow to catch on to. More than a shuffler of abstract digits, the Universal Machine simulates physical objects and bears on the very essence of physical reality. David Deutsch, both mathematician and physicist, would eventually bring this to light and allot it the status of a physics principle. In a trenchant paper— fittingly presented to the same scientific society in London Turing had presented his to 47 years before—Deutsch expanded the notion of computer universality to encompass quantum mechanics and laid the theoretical groundwork for a universal computer capable of rendering the world outside.

Technology wasn't far behind. Computers were developed that would do just that. Some are now commercially available. They are still far from achieving the perfect mimicry Deutsch envisioned—they are digital computers, not quantum computers as theory demands. But even so they manage to catch a good likeness of things, including their dynamics. Indeed, a whole industry has sprung up devoted to simulating environments.

The industry's latest nine day's wonder are the "virtual-reality generators," complex machines whose workings are kept tightly under wraps. But despite all the hoopla, those machines hold no great secret. Stripped of their accoutrements, they are all basically the same. The heart of the machine—or should I say, brain?—is a computer programmed to generate images that have enough of the real world in them that we can interact with them as if they were real.

Take the flight simulator, for example. This virtual-reality machine is now widely used for training pilots. Its computer is programmed with data from an aircraft and its surroundings—sounds, weather, air turbulence, airport layout, mountains en route, and so on. That is a wide scale of data, so the computer will generate a rich imagery, engaging several senses at once. It's quite an experience: strapped into the machine, you take off to jolts of acceleration and the roar of engines, while images fly by the windows, the fuselage quivers, instrument panels flicker, and radar blips of mountains or approaching aircraft appear on video screens; and as you move the right levers, the "aircraft" follows your every command and securely wings you from airport to airport (even through a mountain if the system isn't properly programmed with the laws of physics), and finally you land, to a round of applause.

If that's not enough, a stylish architect will offer you another treat. He has a machine whose computer is programmed to generate a three-dimensional image from a standard architect's drawing, and it comes equipped with projection gear and a sense-stimulating helmet straight out of Aldous Huxley's *Brave New World*. That machine allows you to view the prospective house from all angles; palm it; walk through it; experiment with its lighting, acoustics, and furniture layout; adjust its heating and cooling to the weather; and more.

However, the pyrotechnics should not distract us from the heart of the matter: the computer. It is in there, not in the accessory equipment, that the actual generation of the virtual image takes place. We might even say that the image resides there, as in a sense it does, for *the molecular structure of the computer hardware, the momentary molecular configuration, contains the mathematical transform of the information from the world outside.* And this transform is the quintessence of that information, its molecular embodiment, which could be staged in any number of ways. You could dispense with the accessories that do the staging—even with the observer—but never in all your born days with the computer.

But bear in mind, those machine simulations are not perfect. Digital computers cannot render *all* of the world. They are continuous systems, and hence cannot render the discontinuous quantal world; only a quantum computer can.* But they catch enough of a likeness. And that is enough for some purposes of our brain.

The Neuronal Virtual-Reality Generator

So the question now before us is, what kind of computer is the brain? What is the operational mode of this convoluted, bewilderingly tangled web of neurons that has so much in common with a virtual-reality machine?

Well, it may be too early to say what sort of computer it is, but we may say what it is not—or more assuredly, what it does not. It does not render the outside world precisely or completely. It is simply not designed for that—if it is at all admissible to speak of a design in the case of a product of Evolution. We are dealing here with a natural computer that was shaped by evolutionary pressures. Those pressures go back to the loop/trellis transition—and the only thing such pressures have regard for is organismic survival, not mathematics.

This could be said of every piece of computer hardware up there—every convolution, every crinkle of that precious grey matter was kneaded by those pressures. And even had this computer been designed, it couldn't have been more single-mindedly, for the pressures relentlessly propelled the development to one end: the computing of future events.

---

*It is this type of computer, the quantum computer, for which the above-mentioned physics principle holds. A quantum computer is similar to a digital one in its input and output, but its inner workings and logic are very different. It uses the quantum states inside ions and molecules for its operation, which are immensely more numerous than those of the digital counterpart and make it immensely more powerful (we will deal with that further on). Formally, however, it is still a Universal Turing Machine—one that uses a qubit, instead of a bit, in each square of the tape and has quantum states, instead of macroscopic digital states, in its "head." It is for such a computer that Deutsch formulated the strong version of the physics principle, to wit, "*Every finitely realizable physical system can be perfectly simulated by a universal model computing machine operating by finite means.*"

As for the web's virtual-reality operation, that, in an evolutionary sense, is but a sideline. It goes hand in hand with the primary computing and uses the same hardware and data bank. Both operations grew in the course of time under the same evolutionary pressures and, as a result, more and more of the outside world was rendered. Eventually the rendering would become important in its own right, so much so that in the end it came to command the conscious scene.

Here the question may intrude, where in the brain is the scene? Where is the screen on which the brain stages the images of the world outside? Well, that's the wrong sort of question. It conjures up Descartes's Homunculus and other pesky ghosts. We have no use for those. We'll take our cues from the virtual-reality machines wherein the world outside is represented by the momentary state of the computer hardware. This state is a transform of the information from the world outside and is sufficient by itself for virtual imagery. This has profound implications for our perception of reality, which we deal with in detail further on. Meanwhile, let's say that in the brain computer, the information of the world outside, its pertinent information transforms, resides in the momentary fine-structure of the neuron trellis. With its trillions of cells, that network offers astronomical configurational possibilities—enough to represent the world.

Our Biased World Picture

The question now is, what of the world is represented by the network? Obviously there is much more out there than the information that our senses channel into that gray web. But if it operates like a Universal Turing Machine, there are in principle no limits to representation. The "out there," together with the web's peripheral sensory lines, would constitute the input stretch of the Turing tape. Think of it as a very long tape with a large number of digitized squares (the digital electrical nerve signals) running through a "head" ruled by a program. Given a large enough memory, such a computer would be able to render any pattern.

The limits here are set at the brain's periphery by its sensory cells, not by its centers. These cells respond to a narrow energy range. Take our visual pictures. Those derive from a rich information input—indeed, there is none richer for the human brain. Yet it is only a fraction of the information that is extant; the input is limited by the rhodopsin demons to a narrow window of 300–800 nanometer wavelength. No doubt there is much more information out there; we only have to use artificial sensors widening that window to other wavelengths, say, infrared, and all sorts of formerly hidden things heave into sight.

What is normally rendered by the brain is thus an incomplete picture. It is a picture with a bias etched in by age-old evolutionary pressures for survival. Every fiber of the brain computer, every piece of hardware, carries the bias. This makes it quite another thing from the technological computers, as it is not designed to render a faithful world picture, but a useful one—useful for the survival and well-being of the organism.

That bias also applies to the rendering of the time dimension. We have already seen how our world picture is distorted by the sensory-demon tandems standing guard at the brain periphery, as the information about slow environmental events is systematically filtered out (see figure 3.6). That sort of filtering, however, may start even before the environmental information hits the tandems. A case in point is the Pacinian corpuscle, a ubiquitous little sense organ providing information about mechanical vibrations in skin, tendons, muscle sheaths, mesentery, and so on. This sense organ can cut down the information flow even before it hits the sensory dendrite, pushing the filtering to the very limit: two thousandths of a second, regardless of the duration of the environmental mechanical event. We'll see further on how the sense organ does it—it is one of Evolution's amazing engineering feats. Here let me just say, as far as this particular information channel of our brain is concerned, the world might as well stand still after that split second.*

One could hardly think of a more blatant twisting of reality. And that's not all. In this and other sensory-information channels, systematic

---

*There are, however, information channels running in parallel, which are served by sense organs whose filter action of slow events is not as severe.

distortions occur higher up in the brain, at the synaptic stations along the way to the cortex. There, contrasts are accented, contours sharpened, information about things that move favored over information about things that don't, and so on. It is as if the brain would systematically give the spin to our world picture.

An engineer may take a dim view of that. It's hardly the way she designs her virtual-reality machines. And it's certainly not what a mathematician would like. But then, he does mathematics for the platonic joy of it, not at the behest of Lady Evolution. To him, the difference between light and shade, or between a thing that moves and one that doesn't, is but a footnote.

Computing by Neurons

But spin and all, that gray web is a computer. Every neuron in there does computations. And it's not merely an operation in which one variable is transformed into another—that sort of thing any abacus or slide rule can do—but an operation that serves biologically meaningful functions.

Such operations start the moment the environmental information is encoded by the sensory-demon tandems. The generator potentials there and the digital electrical signals are the information transforms (chapters 3 and 4), the first of a long series of transforms—or analogs, in computer terms—of the environmental information. Each analog involves computations, which in the case of the generator potential are by and large nonlinear multiplications and, in the case of the digital signals, simple linear ones.

That run of things continues higher up, as the information is relayed at synaptic stations in the brain. Like the sensory periphery, much of the computing up there takes place at dendrites. However, unlike their peripheral counterparts, these dendrites often branch—many of them typically form extensive trees. This makes the computations more complex, especially when those dendrites receive, as they often do, both excitatory and inhibitory inputs (see figure 12.2). The analog, namely the change in dendritic membrane voltage, is distributed over the membrane. That requires no extra energy or extra information. The distribution is passive;

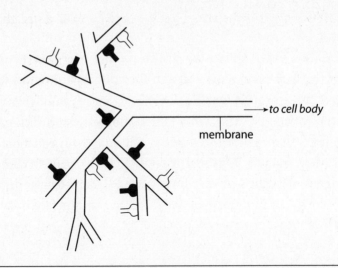

to cell body

membrane

Figure 12.2. A Dendritic Tree with Excitatory and Inhibitory Information Input. Excitatory presynaptic axon endings are shown in white; inhibitory ones in black. The resulting changes in membrane voltage spread passively over the membrane's capacitative and resistive network, much as in a branched electric cable.

the voltage change spreads over the capacitative and resistive components of the membrane. It is not unlike the situation at peripheral sensory dendrites where the generator potentials are produced (figure 4.2), except that here we are dealing with a whole dendritic tree, a branched electric cable instead of a straight one.

Adding to the complexity are the multiple information inputs. Figure 12.2 sketches a situation in which excitatory and inhibitory (GABA-releasing) axon endings impinge on the dendritic tree—a typical situation in the brain of vertebrates. The excitatory input here causes the postsynaptic membrane potential to decrease, while the inhibitory input counteracts that by opening up chloride channels that shunt the membrane. That makes for a rich ground of nonlinear operations, which the biophysicist Christof Koch and his colleagues have elegantly analyzed.

Simpler computations take place at dendritic trees where the inhibitory input, rather than shunting the membrane, actually causes the postsynaptic membrane potential to rise—just the opposite effect of the excitatory input. Then the analog computation is a straightforward subtraction.*

That gives us an idea of the sort of computations the brain uses to extract biologically useful information from the data input of the world outside. But who pays for that? Who pays the thermodynamic piper? Each computation represents an information gain, and such gains have their thermodynamic price. Indeed, as the cognition equation (chapter 2) tells us at a glance, the irreducible cost of that universe of computation and cognition is astronomical. But the number of molecular demons stationed along the dendritic trees and other synaptic surfaces is astronomical, too, and we have seen what thermodynamic exploits their likes are capable of (chapter 2).

And if we get right down to it, all these protein demons perform a computation as part of their regular cognition cycles. Consider, for instance, the action of the relatively simple protein demon in figure 2.6 shortly before he recognizes the molecule of his choice. He starts his cycle with a computation, an integration: he compares the contours of the potential mate with those of his own cavity, and when a certain proportion of the contours match on summation, he says, "Aha, you are my type!"

This "aha" may not be much by Archimedean standards, but that's beside the point. The underlying operation is a computation, and the little demon goes it all alone. Indeed, he ranks higher in self-sufficiency than your home computer. The information states of the protein demon are embodied by distinct molecular configurations, just as the information states of the home computer (its "logical states") are embodied by distinct configurations of the hardware. But whereas the demon procures

---

*In both cases cited, γ-aminobutyric acid (GABA) is the transmitter of the inhibitory input, but the receptors for the transmitter on the dendritic trees are different. In the first case, it is a $GABA_A$-receptor, which opens up chloride channels, shunting the membrane and causing the membrane potential to stay close to its resting value; whereas in the second case, the receptor is $GABA_B$, which opens potassium channels, causing the membrane potential to rise. The inhibitory effect in the $GABA_B$ case amounts to an overpolarizing battery, which linearizes the computation and, in the limit, the computation becomes a simple subtraction.

the extrinsic information for the resetting of his original configuration by himself, the computer needs the assistance of a human operator. It is precisely during the resetting (which in the case of the home computer corresponds to the clearing of the memory register) that the thermodynamic account is settled and the irreducible information price is paid, as demanded by the Second Law.*

That price, we have seen, the molecular demons pay punctiliously. They may not be on the side of the angels, but they always keep on the right side of the law.

Correcting Our World Picture

Let's return now to the brain computer as a whole and take another look at that obstinate bias. The bias is as old as Methuselah and imbues hardware, software, virtual imagery, everything. But is that imagery really so irremediably misbegotten as a first reading of the data input and its initial processings suggests? The input, it is true, is systematically distorted. But those distortions are not carved in stone. They are distortions of information, largely owing to nonlinearities in the sensory signal encoding, and such distortions are malleable. So if those nonlinearities have bent the original information out of shape, it can be unbent again.

Indeed, there is nothing in the laws of physics standing in the way of such correction. Mathematically, an informational distortion amounts to a fixed operation that can be undone by the inverse operation. For engineers designing telephone, radio, and TV equipment, that is par for the course. They integrate the necessary computer hardware with the encoder hardware. That takes up apparatus space—in our TV sets, it clutters up much of the inside—but without it, without signal correction, we wouldn't stand a chance of recognizing a TV picture beamed from afar.

---

*There is an additional thermodynamics price to be paid for the thermal jitter of the molecular structure of the demon, but this is an incidental cost of cognition rather than a fundamentally irreducible one, as in the configurational (allosteric) resetting.

Signal correction runs into money, and it makes no difference whether they are radio waves or electrical signals in telephone cables or in nerves. Every signal corrected means an information gain, and you can measure it in bits. But you need to pay for them, even if, as such operations go, it may seem like paying through the nose. In practice, it means supplying the computer system with lots of information.

Thus our questions become: Where in the brain does that information come from, and is there enough of it on hand to correct for the nonlinearities and other distortions of the sensory input? More specifically, is there enough information for straightening out the distorted virtual images? Well, we needn't look very far: there are huge amounts of information stocked up in the trellis, which over the long haul was laid in store for the purposes of the aboriginal trellis function, forecognition. Most of that is recurrent information that the world outside so bountifully supplies and gets used for computing the odds of future recurrences. However, there is no reason that information couldn't also be used for correcting the distorted sensory virtual images; the required operations are not unduly complex. They are standard fare in modern imaging technology—thanks to them pictures beamed by probes from Mars or spy satellites are uncannily clear. So it would not be farfetched to think that a web, like the brain's, with more than seven orders of magnitude more units than today's most powerful technological computer, is up to such a corrective task.

Flying the Coop of Our Senses

Such corrective actions are but a few operational steps away from a more creative pursuit: *combining stored information pieces, namely, pieces of recurrent information, and computing the reciprocal relations between them.* Such operations go well beyond the mere ex post facto straightening out of virtual-imagery distortions; they provide a perspicuity of their own, greatly widening our cognition of external reality. Indeed, given the vast amounts of information our memory holds in store, the number of combinatorial possibilities is immense, offering a never-ending source of inventiveness.

That capability developed only recently in evolution, probably no more than 40,000 years ago*—it is the latest stage in the development of the neuron trellis. Late as it came, it started a whole new ball game: the players were no longer hemmed in by narrow sensory constraints, but could now wing themselves above the sensory horizon and see things formerly hidden from them. Such seeing is what, perhaps more than anything, marks us as humans. And I don't mean here a seeing assisted by artificial sensors—that sort of expansion of the sensory horizon has little need for combining and correlating information pieces—but a true transcending of the sensory sphere, a becoming aware of a deeper reality.

I use the word *transcending* to signal that we are entering philosophers' territory—Kant called it "pure reason." But we need not walk on tiptoes here; we are entering through the front gate, the information gate, a vantage philosophers never had. Anyway, it's no one's private territory. It has been a common ground for philosophers and scientists for over 400 years, although they operated in different landscapes. And whether we call it "pure reason," "the mind's eye," or "computing the unseen," or what have you, we mean the same thing. We mean what the brain does by virtue of its structure as a logical system.

Let me give some examples drawn from science. Good examples are certainly not hard to come by; the searching for relations between recurring information pieces is precisely the scientist's game.

One example comes from physics and concerns the discovery of the first quantum particle, the electron. This took place in the 1890s. At that time physicists were puzzling over the mysterious rays that were emit-

---

*That was the time when our Cro-Magnon ancestors supplanted the Neanderthals, a time of unprecedented inventiveness, according to the fossil record. Ingenious instruments made of antler and bone—needles, awls, fish hooks—then made their debut. And so did weapons designed to kill from a distance—harpoons, spears, bows, and arrows. It was the time of fired-clay sculptures, bas reliefs, polychrome paintings, including abstract ones—a juncture in hominid development Jared Diamond called the Great Leap Forward. The fossil record of earlier hominids shows no comparable inventiveness and premeditation. Their brain wasn't much larger five to seven million years ago than that of today's primates. Even three million years ago the hominid brain (e.g., *Australopithecus africanus*) was only about a third the size of ours. Only about 1.7 million years ago would it reach a comparable size (*Homo erectus*).

ted in the vacuum from a metal plate when it was connected to the negative side of a high-voltage source (hence the name cathode rays). The rays could be seen—or rather their effect could—as they hit a phosphorescent screen, which would then emit light. Two men investigated the nature of those rays, subjecting them to electric and magnetic fields: Joseph John Thompson at the Physics Laboratory of Cambridge and Walter Kaufmann at Berlin. Both measured the rays' electric charge and mass, with essentially the same results. But only the one who made the most of the recurring relationship between the information pieces saw what the rays are made of; Thompson, from the charge/mass ratio, cogently concluded that the rays are made of a novel particle, the electron, a particle many times smaller than the smallest atom. Not that Kaufmann's measurements weren't accurate—in fact, if anything, they were more accurate than his competitor's. But all he did was report the rays' electric charge and mass, whereas Thompson, with the same raw data and but a few more operational brain steps, saw a deeper reality: the existence of a particle, more elementary than anyone had seen before.

The atom had been discovered in like manner 2,000 years before. Then it was the philosopher and physicist Democritus of Abdera who used the relationship between pertinent information pieces to draw the crucial inference. By modern standards, those pieces weren't much. But the point is that, by making the most of what was available at the time, he saw beyond anybody's eyes. Indeed, he unveiled a whole world picture in which all things have an unseen grain.

It's not just physicists who make forward leaps that way. All scientists do. The laws of nature are in every thing. It just takes a discerning mind to wrest them out. In biology, such wresting has not yet advanced to the degree it has in physics. Biology is younger, and a good deal of effort still goes into identifying and extracting the significant pieces of information from the pile. Nevertheless, where those preliminaries are over and done with, the crucial leaps are made the same way, with the mind's eye: by establishing the reciprocal relationships between the pieces.

Take our knowledge of the genes. This involved three leaps. The first happened in a little monastery in Brno, in 1865, where Gregor Mendel discovered the laws of heredity. He wrested them out of nothing more than the shapes of peas growing in the garden, that is, from the relationships between the shapes of successive pea generations.

The second leap is less known. It happened 18 years later at Halle University, where the embryologist Wilhelm Roux, pondering the peculiar alignment of chromatin pieces during cell division, concluded that they were the carriers of the units of heredity. Chromatin, we now know, contains the chromosomes, which in turn contain the DNA (deoxyribonucleic acid). Roux saw neither—microscopes then didn't have the necessary resolution to see chromosomes, and DNA was still unknown. Yet making the most of the scant information on hand, namely the recurring spatial relationships between chromatin pieces during cell division, he inferred that they contained qualitatively distinct particles, the units of heredity, whose transmission required a particular alignment of the chromatin pieces.

The third leap was in 1953, and it happened in the same laboratory where Thompson had discovered the electron. It was not by chance that it happened there; the Physics Laboratory at Cambridge had become the hub for probings of molecular structure by X-ray diffraction, and the work there had pushed the resolution for "seeing" things far beyond the limit of resolution of optical microscopes. There, in 1912, William and Lawrence Bragg (father and son) had succeeded in resolving the atomic structure of table-salt crystals—they computed the exact arrangement of the sodium and chloride atoms in the crystals from the X-ray diffraction patterns. That was still a far cry from computing the complex arrangements of atoms in biological macromolecules, of course, but in the 1930s and 1940s a considerable amount of expertise had been assembled in the lab, and there was excitement in the air and momentum to tackle such daunting tasks. It was in that atmosphere that Watson and Crick attempted to solve the structure of DNA.

The rest is history; it was the thunderbolt of our time. But what I wish to point out is that they arrived at their goal by just such a leap of mind as those delineated above: by using the recurrent relationships between information pieces, namely, the spatial relationships between the DNA crystallographic space groups. The crucial things here were not the X-ray information pieces by themselves—those were first in the hands of Rosalind Franklin at Kings College London—but the reciprocal relationships.*

All this goes to show that the practitioners of science have, indeed, learned to wing themselves above the narrow sensory horizon. Not that there is no longer room for the senses in the scientific endeavor; that's unlikely to ever happen. Too much ties us to them: memories, emotions, all of our sensory brain. There will always be a place for the sensory in science, if only to hold up the mirror to reality. That is the test all science pictures ultimately must stand—those pictures are no castles in the sky. And whether for this purpose the senses are used directly or indirectly through instruments, it's all the same—a scientist may only fly so high.

However, there are moments when scientists come pretty close to completely flying the sensory coop and making discoveries with the mind's eye. Such moments are rare and, for the time being, confined to physics. For an example I turn once more to the story of the electron and pick up the thread where we left it, 30 years after Thompson's discovery.

It was then that Paul Dirac (1928), merging quantum and relativity theory, wrote his famous wave equation, one of the great achievements in physics. The mathematics here were all-comprehensive; they told us all that there is to tell about the electron—its behavior in electric and magnetic fields, every interaction it may undergo. Moreover—and this is what I am driving at—they predicted the existence of something nobody had seen before: they showed how and why an electron nearing the speed of light must spin around twice before it may show the same face

---

*The discovery here actually involved a triple leap: one starting from the spatial relations of the crystallographic space groups; another from the corresponding helical constraints; and yet another from the whole molecular structure. By the first, Crick and Watson arrived at a structure of two very similar molecular chains running in opposite directions; by the second, at the mode by which the chain's nitrogenous bases were joined together—the only possible mode they could be by hydrogen bonding, without distorting the helical-chain configurations; and by the third, at the mode by which the two chains get duplicated, how like begets like.

once more, and they predicted the existence of a particle with opposite spin and charge (antiparticle), which, contrary to anything in our sensory experience, could suddenly appear out of nothing and vanish into nothing again. And five years later the predicted particle would turn up in experiments, spin, charge, and all!

CHAPTER THIRTEEN

Expanded Reality

A Fine Bouquet

In the early 1930s, on one of his visits to CalTech, Einstein was taken by the astronomer Edwin Hubble up to nearby Mount Wilson, the site of what was then the largest telescope. Hubble had made his famous discovery of the spiral galaxies there, and before long he and his colleagues found that the galaxies were fast drifting apart. This showed that the universe was expanding, confirming a prediction Einstein had made a decade before. Einstein had come to Mount Wilson with his wife, Elsa, and was happily clambering up the telescope's framework, much to the consternation of his host. Meanwhile, Elsa was staying down on the floor, asking questions about the giant mirror of the telescope. Told that it was used to measure the expansion of the universe, she replied: "Well, that sort of thing my husband does on the back of an old envelope."

This was not just wifely pride, but a testimonial to the power of the mind's eye. Einstein's prediction stemmed from his gravitational field equations, the early solutions of which implied an expansion (or a contraction) of the universe. The equations were the centerpiece of his relativity theory, and he worked them out in the period 1905–1915. But the first intimations of this theory he had some 10 years earlier while still in high school. That is when he had the first glimmers born of sensory imagery—or in his own words, when he gave his imagination free rein "to run behind a light wave with the velocity of light."*

---

*From a letter Einstein wrote in 1899 to his friend Mileva Marić, two or three years after the episode (see references).

But the when and where are not important here; it might just as well have been another time or another circumstance that sparked off the critical "seeing." Nature's truths are in every thing. Newton saw the force behind the motions of the planets in the falling apple, and Darwin saw the force behind life's evolution in the activities of the pigeon breeders in his neighborhood. What matters is the seeing of a reality that lies beyond the senses—a deeper reality.

Mathematics and Reality

But why is that deeper reality expressed so often in mathematical form? The answer depends on whom you ask. A mathematician or a physicist is likely to say that nature is mathematical herself—that nature is the supreme mathematician. A neuroscientist will say that mathematics is inherent in the computer structure of the brain—that the perception that the world outside is mathematical is but a fond illusion.

The horns of this dilemma stretch back to Plato and, even before, to Pythagoras. In their time, the question of what constitutes reality already had been the subject of wistful debate. Plato himself had made it the focal point of his *Republic,* dramatizing it in a famous metaphor, the Allegory of the Cave. Inside a cave, and with an eerie, flickering fire illuminating the scene, unenlightened men are chained to a world of shadows—a world where what they see through their senses is only the shadow of reality.

It's the abode of the damned, mathematician style. And there is only one way out of there: through the intellect, the mind's eye. But the great philosopher doesn't leave us dangling by a metaphor. In his *Phaedo* he spells out what his idea of the real world is: a world populated by mathematical entities—straight lines, planes, triangles, squares, circles, spheres, cubes, dodecahedra, and so on. These were timeless mathematical forms that did not depend on humans but existed on their own—forms defined by immutable mathematical relationships.

If you think that notion is but a relic of antiquity, just listen to today's physicists describing *their* reality. They sound hardly less Platonic or less sanguine. Take, for example, Superstring Theory, the latest unification effort joining quantum mechanics and general relativity. Its proponents envision a universe made of tiny filaments of energy—the *strings*—which are some hundred billion times smaller than an atomic nucleus. In this theory, those strings are the ultimate building blocks of energy and matter; vibrating in different dimensional patterns, they give rise to the various species of quantum particles: the quarks, the electrons, the neutrinos, and so on. Such building blocks are the very paragon of Platonic form and represent a reality that could hardly be more unchained from our senses: a reality with 10 dimensions—nine space and one time dimension—six more than we can see or feel.

If anything, Platonism has come into its own in our time, and that development has been going strong for quite some time. Galileo foreshadowed it five centuries ago: "The language that Nature speaks is mathematics," he declared. And physicists have made it their mantra—mathematics became their language for describing and predicting reality.

But what of that other language, the one ordinary human beings speak? Well, don't expect too much there. Among physicists that sort of thing doesn't rate very high. At best, it is accorded a role in the preliminaries to mathematical formulation, the sensory imagery assisting the mind's eye. But once the formulation is done, words are only in the way. Ordinary language thus at most gets assigned the status of a helpful metaphor.

But that knife, I am afraid, can cut two ways. The poet Samuel Coleridge used to attend lectures on science at the Royal Institute of London. Asked by surprised friends why he would put himself through such torment, he replied, "To improve my stock of metaphors."

The Neuron Circuitry of Language

Let's consider the answer of the neuroscientist a little more. He brings to the table three decades of experience with the logic of language plus a century of probings of the brain. From his perspective, mathematics and ordinary language are of a piece and ought to be weighed on the same scale.

A good scale here is that of information; indeed, it's a natural in matters of the brain. We see then at once that mathematics packs vast amounts of information, orders of magnitude more than ordinary language does. But let's first consider ordinary language for its own sake. Although the density isn't so high, ordinary language certainly can carry a good deal of information. Words don't just come randomly out of our mouths, but in a certain order and bear certain relationships with each other: relationships of space, time, hierarchy, etc. We may not be aware of this because the neuronal web sees to it rather reflexively. But grammarians are.

A significant part of the attendant neuronal circuitry lies in the front bulge of our brain. This bulge is of relatively recent evolutionary vintage. It developed some 40,000 years ago in an ancestor of ours; the pertinent neuronal circuitry is on the left side, in a region called Broca's area, after the neurologist who discovered it.* The neuronal circuitry in there shows a special knack for combinatorics. Give it a modest amount of verbal information, say 20 words and 10 choices for each word to begin a sentence, and it will come up with about 10^{20} sentences.

Many such sentences will be beset with ambiguities; they constitute a rather loose language. But our brain has the wherewithal to tighten it by generating auxiliary communication signals: prefixes, suffixes, propositions, participles, pronouns, word endings, and so on. Think of these signals as extra information assets making up for the shortfall in coding stringency. This is why, with a little bit of chatter, we get along so famously.

That notorious bulge in our brain was part of a major expansion of the hominid brain. So it is not unreasonable to assume that other hominids had at least a modicum of language. Primates certainly have. Wild chimpanzees, for example, use some three dozen vocalizations (phonemes) to convey about as many different meanings. However, they cannot string them together in various combinations, as we do with our

---

*This was a landmark discovery in the neurosciences (1863), the first brain area with a specific function to be identified and the first evidence that the two brain hemispheres have different functions. The neurologist Paul Broca discovered the location in a patient who was unable to speak and on postmortem examination turned out to have a tumor there.

phonemes. These strings—we call them words—are a distinctive feature of our species.

We can even go one better and make strings of strings, sentences—and all that as a part of quasi-instinctive combinatorics. Some members of our species even manage to deliver off-the-cuff remarks in long sentences filled with perfectly conjugated dependent clauses, none of them dangling!

The Feathers of the Brain

But what is good for social intercourse is not necessarily good for describing and explaining what goes on beyond the sensory horizon. That calls for a different kind of language, one that is stringently coded, logically self-consistent, and capable of apprehending the reciprocal relationships between things.

Mathematics is such a language. However, the corresponding neuronal circuitry is not so easy to pinpoint as in the case of ordinary language. Mathematics draws more heavily on our consciousness, whose information states occupy vast neuronal spaces in the brain. However, we are not altogether clueless. Clinical neurology lends a hand here; patients with mathematical inabilities (but without obvious memory inability) showed structural deficits in the left frontal bulge of their brain and in the left parietal-occipital region.

This gives us a broad hint. Recall that the frontal bulge is the seat of forecognition, the neuronal-trellis capacity of computing from past experiences the probability of future occurrences (chapter 11). That same capacity thus may well be the basis for higher mathematics, too. This stands to reason because the alternative, a special neuronal circuitry for such mathematics, is unlikely on Darwinian grounds. What selective advantages could the solving of quadratic equations possibly have, or the formulation of topological theorems or wave equations? Whereas it is plausible that those mathematical abilities developed secondarily as by-products or extensions of the life-significant computer capacity of forecognition.

Along this line of thought, we may envision that, from humble beginnings and under the ever-present evolutionary pressures for fast reaction, the neuron trellis grew to a sprangling apparatus for probabilistic computation, eventually becoming co-opted for higher mathematics.

Such co-options are called trans-adaptations in Darwinian terms, and they are by no means oddities. They are among Evolution's favorite information-conserving ploys. The feathers of birds are a classical example: although today the feathers are used for flight and are the quintessential instruments for it, for millions of years, before any creature flew, they served to maintain body temperature. And they are as successful in the secondary adaptation as they were in the primary one. So, putting it briefly, mathematics are the feathers of the brain.

That may not be an inspiring slogan, but it pretty much sums things up.

Mathematics and Forecognition

It is easy to lose sight of the neuronal roots of mathematics when one looks at what mathematicians produce in their ivory towers. But when mathematics is used for more worldly endeavors, like the natural sciences, its kinship with forecognition is clear enough. Consider how laws are discovered in physics: the practitioner searches for recurrent information in the universe, abstracts a rule from it, an algorithm, and then uses the algorithm for prediction. This is how Newton came by his $F = ma$ and Einstein by his $E = mc^2$.

At the center of such discovery is information processing wherein knowledge is generated in the face of uncertainty. Or perhaps I should say "expanded" rather than "generated," to emphasize that the processing transforms extant knowledge, namely, recurring information, into statistical knowledge about the future. And without unduly straining things, that will also cover processes like derivation and induction.[*]

---

[*]Such processings may be defined by a transition function (Γ) of the information states (S) of the outside world before and after the predictions. Thus, as time (t) goes forward, $S(t + 1)$ is given by $\Gamma[S(t), O(t)]$, where $O(t)$ is the output of the brain process at time t and $S(t)$, the outside information state at time t. Held implicit here is that $S(t + 1)$ is the same regardless of $O(t)$; that is, the evolutionary course of the states of the outside world is not significantly perturbed by the brain.

The information processing in science is more involved and more laborious than in forecognition. Scientific laws are hard-earned. Many a measurement may go into the procurement of recurrent information, and more often than not the scientist has to burn the midnight oil to get analytical solutions to differential equations. But the fundamentals are the same as in forecognition.

The Reluctant Sensory Brain

The cognitive expansions that mathematics has wrought are vast—they laid open a whole new reality. But they have had a curious, one might say perverse, psychological effect: the more the knowledge is expanded, the less real it seems to us. This holds true for much of modern physics. We need look no further than Newton's $F = ma$ and Einstein's $E = mc^2$: Newton's algorithm contains but a fraction of the information of Einstein's, yet it strikes us as being much closer to reality. That a force (F) acting on a given mass (m) should cause it to move with an acceleration (a) proportional to the force seems only natural to us. It gibes with our daily sensory experience; we throw a stone with more force, and it moves visibly faster than when we throw it with less force. We also intuitively grasp that it potentially has more energy; it visibly and palpably causes more damage. On the other hand, that it should still contain energy after it has stopped, as Einstein's algorithm tells us, strikes us as unreal. Yet it *is* real. Atom bombs, nuclear power stations, and particle accelerators leave no doubt about it.

Then why is that truth so hard to swallow? It certainly couldn't be because of abstruse mathematics. Einstein's algorithm, indeed, the whole condensation process whereby he arrived at it, is but high school algebra. Nor could the reason be that the algorithm doesn't encompass the cognitive processes in our brain. On the contrary, with each cognitive act a bit of organic mass must bite the dust, as information changes carrier. And it's happening in you right now as you read this line: the rhodopsin in your eyes is bereaved of a bit of mass, as the energy in the photons carrying the information from the page is converted into electricity by neurons. And so on and on, each process, each macromolecular demon, along the way from eye to cerebral cortex, makes his bow to $E = mc^2$.

The reason for our recalcitrance lies elsewhere. It lies in the primeval machinery of forecognition of our sensory brain. That machinery is a special-purpose computer designed to operate with information to further organismic survival and well-being: life-significant information. And here is where the shoe pinches the toe, or better said, doesn't pinch it. The changes in mass above, the Δms, are all very, very small. They are real, but by no stretch of imagination are they life-significant.

So if there is a culprit in this story, it is the ancient forecognition machinery. There is no helping it, it will rear its petty head against reason; it has done so historically against all major expansions of our sensory sphere wrought by mathematics. A notorious case is quantum theory. It doesn't help to stand on dignity here or fob off one's feeling of unreality on something else. It's our sensory brain that is speaking then, and with its age-old bias geared to what is useful for life, it makes a poor judge of reality.

The Limits of Knowledge

Having touted the virtues, I must now tell about the limitations of mathematics. Although mathematics is good for apprehending the orderly parts of the physics universe, it is not for apprehending the chaotic parts—and this is true even when they are above thermodynamic equilibrium and contain information.

Consider the turbulences in a thunderstorm, or in a whirlpool, or in a kettle of boiling water, or the tumultuous electrical activity of a heart gone amok. All these are chaotic systems that contain considerable amounts of information. Yet try as you may, you cannot apprehend enough of them with mathematics and predict the behavior of these systems—and even the best of mathematicians must eat humble pie here.

If we look at the particle ground floor of these systems, we see why. They are highly nonlinear systems, and the innate randomness of their quantum particles is amplified by the nonlinearities. Indeed, so enormous is the amplification that even the macroscopic manifestations become random here. That is what puts these systems beyond the apprehension of mathematics; the nonlinearities amplify the inherent uncertainties in the system at an exponential rate, always putting the system a step beyond the reach of mathematical prediction—or put in terms of information, the uncertainties in the system grow faster than the capacity of information processing.

Mathematics is just as powerless in the opposite direction, retrodiction (analysis). The motions in the chaotic system are so complex and the trajectories so varied that no observation, whatever its precision, could give us the exact initial condition from which we could derive the system's dynamic behavior.

We reach here the limits of human knowledge.*

A Note about Reality

All things considered, our grasp of the reality of the world outside, although vastly expanded by mathematics, is but a partial grasp. And saddled as we are with a brain wired in our distant past for lowbrow computations, it may well be a partial grasp forever.

Or am I being too hasty in my summation? There is after all a notorious character from Wonderland, whose brain seemed to have been wired otherwise:

"I see nobody on the road," said Alice.
"I only wish I had such eyes," the King remarked in a fretful tone,
"To be able to see nobody! And at that distance too!"

---

*Chaos is not the only forbidden fruit on the tree of knowledge. The ultimate epistemological limits lie in mathematics itself, as shown by the incompleteness theorems of Kurt Gödel and Gregory Chaitin.

CHAPTER FOURTEEN

Information Processing in the Brain

After our excursion into expanded reality, we return to the primal sensory reality, the one bred into the bone, so to speak. Picking up the thread of our story in chapter 12, we'll follow the information stream from the senses to the cerebral cortex to track down how the web of our sensory brain generates its virtual images of the world outside.

Let's begin with a synopsis of what lies ahead. Those virtual images are of a very different sort from the ones our technological virtual-reality computers generate. They are not bit-by-bit representations of a three-dimensional world—Cartesian pictures, perish the thought!—but higher-order information transforms. The transforms do not match the optical image on the retina pixel-by-pixel, but are highly abstract forms of that image. It is not that the eye lacks resolution; on the contrary, the sensory grain is reasonably fine. There are about a hundred million sensory cells (rods and cones) in the human retina, and at that level the image of the world outside is an array of regularly spaced points at which there is a certain amount of light and color composition. However, that sort of information doesn't get to the cortex. What gets there are information transforms of that punctiform image, transforms in which each point is represented in a specific life-significant context in various parts of the cortex. For instance, the angle of orientation of things is represented in one part, while color composition is in another. Think of those transforms as highly abstract versions of the original image, which become more and more abstract as the information ascends to and through the cortex. However, despite the serial abstractions and the wide dispersion of information, the original information doesn't get scrambled—the abstraction and dispersion go by strict rules, and the dispersed information is processed in parallel.

Cell Organization in the Brain

The brain contains 10^{12} cells. Each cell may receive information from hundreds or thousands of other cells and in turn transmit information to hundreds or thousands of others. The total number of synaptic connections in the white and gray should therefore be somewhere around 10^{14} or 10^{15}.

Yet despite the astronomical number of axons and connections, there is order in that web. The great neuroanatomist Santiago Ramón y Cajal bequeathed us this knowledge at the turn of the twentieth century. His studies on silver-stained slices of the brain laid the foundations for brain physiology. I reproduce a drawing of his, a microscopic view of an outcropping of the brain, the retina. It speaks for itself, and even after a hundred years there is no need to add a word to it.

The order in the brain is governed by a simple organizational principle: neurons with similar or related functions are grouped together and interconnected. That gets their messages to each other fast and takes up little room. Think of it as a computer-wiring problem: there are about 80,000 miles of wiring to be packed

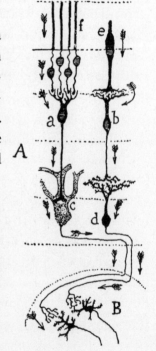

Figure 14.1. A cross section through the vertebrate retina showing the nerve cells arrayed in it and the photo-receptors. This diagram was drawn by Ramón y Cajal about 1911 from a silver-stained section (Golgi method). (a, b) bipolar cells, (c, d) ganglion cells with axons forming the optic nerve; (e) cone; (f) rods; (A) retina; (B) geniculate nucleus. The arrows indicate the direction of the information flow. (From Ramón y Cajal 1911.)

into the human skull, an inextensible space of about 1,500 cubic centimeters. Positioning the information-processing elements with related functions close together for interconnection is the efficient solution.

In the cortex the information-processing units are tightly packed; there are 100,000 neurons per cubic millimeter of cortex (see figure 14.2). In the areas receiving the information ascending from the senses (*primary sensory cortex*), neurons coding for related features of things outside—for example, orientation, depth, direction of movement, color— are stacked above each other, forming columns of about half a millimeter in diameter. This local organization holds the key to spatial coding in the cortex. In a sense, the columns parcel up the outside world into biological domains of representation—think of them as computer files with a practical bias.

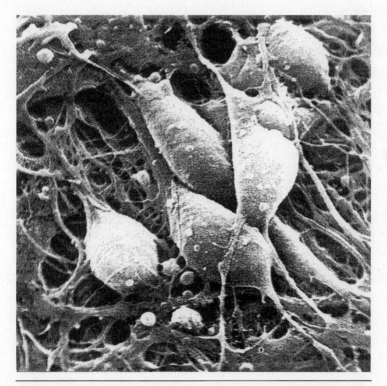

Figure 14.2. The Neuron Web in the Cortex. An electron-microscope picture showing the cell bodies and branching dendrites and axons enlarged about 500 times. In that dense web, each neuron may receive electrical signals from hundreds or thousands of dendrites or axons.

Together with their entourage of dendrites and axons, the cortical neurons form a complex web. There are about a billion synapses per cubic millimeter of cortex (a handy rule of thumb is 1 synapse/μm^3). The synapses are for communications within and outside the cortex. The majority are for the former—the cortex mostly talks to itself.

There also is a fair degree of horizontal order in the cortex; the cells are arrayed in layers. The topmost layer contains mostly pathways providing feedback from other cortical regions. The other layers contain the cell bodies of the cortical neurons plus short axons running up and down the columns (see figure 14.3).

The long-distance communications are handled by cells in the lower two layers, the *pyramidal cells*. This is the predominant cortical cell type; three out of every four cells are pyramidal ones. The dendrites of these cells rise straight to the surface of the cortex, and their axons go to other, often faraway, cortical regions; as the axons relay their messages to the distant targets, carbon copies are sent to neighboring cells via short branches. Another common cortical neuron type relays messages only to neighbors—one out of every five cortical neurons belongs to this type.*

Cortical Information-Processing Units

The foregoing is but a sketchy picture of the gray mantle of our brain, to help us chart a course through its neuron web. That web is vast and complex. But first and foremost is the question, is it robust to noise?

One could tackle this question in a number of ways, but having recourse to microelectrodes, we'll do so at the unit level of the web, that is, the level of the cortical cell at a node where sensory information gets processed. This cuts to the chase, as the question becomes: will the cell process incoming sensory information reproducibly despite local noise?

---

*The axons of these neurons release gamma-amino butyric acid (GABA), an inhibitory neurotransmitter, at their terminals. The axon terminals of the pyramidal cells release glutamate, an excitatory neurotransmitter.

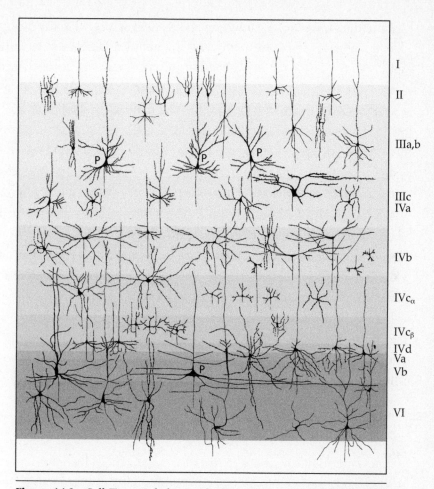

Figure 14.3. Cell Types Inhabiting the Human Primary Visual Cortex. A section through the cortex showing the various cell layers (I–VI; I is the topmost layer immediately below the meningeal membrane). The predominant type are pyramidal cells (*P*); the largest are in layer *Vb* issuing long-branched axons. Camera-lucida drawing of silver-stained cells. (After Braak 1976.)

Consider a pyramidal cell. Deployed at the conflux of incoming information lines, the cell integrates the information from hundreds, if not thousands, of synaptic inputs. Each synaptic input here has only a weak effect on its own, far too weak to elicit a digital output from the cell; the individual effect amounts to only 1–5 percent of the cell's digital firing threshold. Yet given a constant input, the cell will on the average generate a reproducible digital output. It is a dependable information-processing unit.

We may lean here on a large body of experiments carried out on monkeys, whose cortical cells were probed with microelectrodes while the animals performed a defined cognitive task. The digital output of these cortical cells—the average rate of digital electrical pulses integrated over a period of time, $f(t)$—proved to be a dependable predictor of the animals' behavior.

Further assurance comes from experiments on human subjects. It has become possible in recent years to monitor the digital-signal output from cortical cells in subjects who are fully conscious. Among other cortical regions, such signal activity was recorded with microelectrodes in the visual cortex, while the subject was looking at *and seeing* images projected onto a screen. We'll deal with some of these experiments in detail in a later chapter. Here I just state the gist of the results pertaining to the question raised above: the rate of the digital signal output of these cells is a dependable predictor of sensory perception.

Cortical-Cell Topography and Worldview

Our next question concerns the spatial arrangement of the information-processing cortical-cell units. Is their position in the cortical web related to that of the sensory cells that receive the information at the body periphery? Or put in terms of our perception of the world, is that position a predictor of the location in the outside world from which the information originates?

The answer to both is again a reassuring yes, and comes from a variety of sensory systems. Let's start with the system that carries the most weight in our world perception, our sense of vision. The outside world gets imaged on the retina, and the information from that optical image goes to the cortical web of the occipital lobe of the brain (see figure 14.4). The information lines end up on the web in a certain order, though one would hardly suspect it from the helter-skelter way the axons from the retina run in the optic nerve. But by the time they get to the next way sta-

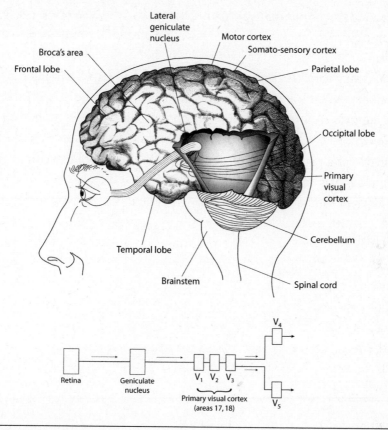

Figure 14.4. The Human Brain. The visual information lines from eye to primary visual cortex are roughed in. *Bottom:* A schematic of the information lines up to areas V4 and V5. The lines go on to the inferior temporal cortex and more distant parts of the cortex, including the frontal lobe (not shown).

tion, the geniculate nucleus, the axons have sorted themselves out, ending up in determinate order on geniculate cells. The geniculate cells then take it from there; their axons, after fanning out through the interior of the brain, end up in an orderly manner on the cells of the cortex.

Beyond the information lines going to the primary visual cortex, the disposition of the lines is just as orderly—the topological game is played methodically over and over in the cortical web. In the end, the retina pixels get systematically mapped on the web's cell grain in various parts of the cortex (figure 14.4, bottom).

The same ordering principle rules the organization in the senses of hearing, touch, and pain. If one tracks the information lines from touch receptors, say, Pacinian corpuscles in the hand, to their cortical destinations, the receptors' skin locations are found to systematically map on the cortex. With Pacinian corpuscles, such tracking is made easy by their axons running singly over a stretch in the skin—nature offers the scientist private sensory information lines on a silver platter here. The corpuscles map systematically on neurons of the way station in the thalamus and on neurons of the web in the parietal cortex (somato-sensory cortex; see figure 14.5).

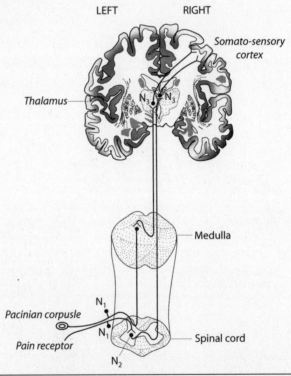

Figure 14.5. The Information Lines of Touch and Pain. The diagram shows the neuronal pathways to the cortex for a touch receptor (Pacinian corpuscle) and a pain receptor in the left arm. (N_1) the corresponding primary sensory neurons in the spinal ganglia; (N_2) secondary neurons in spinal cord or medulla; (N_3) tertiary neurons in the thalamus. The pathway of the touch receptor crosses over to the right side in the medulla and that of the pain receptor, lower down in the spinal cord. Temperature-receptor pathways (not shown) have a crossover pattern similar to that of the latter.

But something may strike you as odd. The information lines cross over to the other side of the body—it is the right side of the brain that feels the left side of our body, and vice versa. And this is not a quirk of the sense of touch; the lines of the senses of temperature and pain also cross over (figure 14.5).

The oblique layout of the sensory lines has ancient evolutionary roots. It is a leftover from a time when the senses mainly served organisms to avoid obstacles or to skedaddle, and the reactions were carried out by muscles on the opposite side of the body.

A Last-Minute Change in Worldview

Let's return to our sense of vision, for there still awaits us a special twist: only about half of the information lines cross over; the other half go to the same side of the brain they originate from (see figure 14.6).

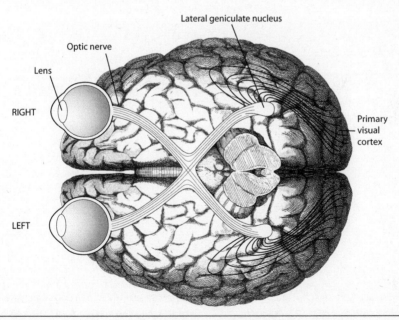

Figure 14.6. A Bottom View of the Brain and Visual Information Lines. About half of the lines coming from one eye cross over to the other side of the brain. The lenses in the eyes reverse the images on the retinas, so that the right side of each retina receives information exclusively from the left half of the visual field, and vice versa.

Or perhaps to place things in historical perspective, I should have said, they *no longer* go to the other side, for this is a rather recent development in our brain and that of a few other creatures sporting frontward eyes.* But why the change in topography? The accumulated wisdom of the ages is that this change expedites frontward vision. But that's hardly an explanation. What possible purpose could the elaborate halving of visual information lines serve when birds, salamanders, frogs, and fish get along famously without it? And elaborate is the right word. If anything, it is an understatement, for there is not just a numerical halving of lines but a meticulous partitioning along pixel borderlines, so that the information lines from the medial retina halves (the halves close to the nose) go to the opposite side of the brain, and those from the lateral retina halves, to the same side (see figure 14.7).

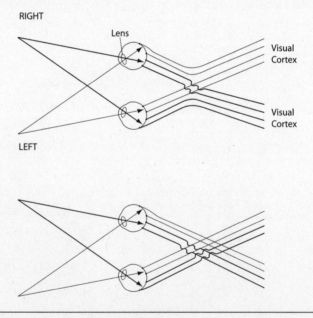

Figure 14.7. *Top:* The new topography, wherein visual-information lines partially cross over, depending on their points of origin on the retina. *Bottom:* The old topography, with all visual information lines crossing over.

*Cats, dogs, and monkeys. Mice, rabbits, and horses show an intermediate development wherein only a small fraction of retina axons no longer cross over.

It's not easy to immediately make head or tails of this change in the layout of information lines, because the lenses invert the images on the retinas. But if we ignore the layout for a moment and fix on the outside locations where the images come from and on the corresponding retina pixels and their cortical mapping, we begin to see through the smoke and mirrors: whereas formerly the pixels on the left retina, regardless of where in the outside world the information came from, mapped on the right side of the brain, and vice versa, now the pixels receiving information from the left side of the outside world, regardless of what retina they are on, map on the right side of the brain, and vice versa.

So more than just a change in line layout and pixel mapping, we have a change in world mapping here—a new outlook on the world.

This is a major development in biological evolution, and what makes it all the more intriguing is that it happened so recently—its footprints are still fresh enough to make out some intermediate steps in other species (see footnote on page 174). Observing the demarche of Evolution from so near gives one an eerie feeling and is a humbling reminder that our brain is but a work in progress.

Retrieving a Lost Dimension

But there still remains the question of why the major brain change occurred. We can make short shrift of the thought that it allowed us to see more of the world outside; birds and frogs have wider angles of view than we have, and as for sharpness, the eagle's eye is a tough act to follow.

No, in all likelihood the change has to do with speed of information processing, namely, information about the depth of things in the world outside. That dimension is lost in the optical imagery of our eyes, and the new layout of sensory lines and brain-cell topography offer a decisive computational advantage for retrieving it.

Let's consider the basis of three-dimensional vision. Experience tells us that things have three dimensions. Nevertheless, it is puzzling that we should *see* the third dimension when all that is pictured on our retinas has only two. Where does that extra dimension spring from? How does the brain obtain depth from pictures flat as pancakes?

The third dimension is real, of course. Its authenticity is confirmed by our senses of touch and hearing, not to mention countless scientific measurements—I wish we could be as sure of some of the other dimensions in modern physics! However, the third dimension gets lost on the retina. But not irretrievably. It can be recovered through computation, and nothing in principle stands in the way of such a recovery, given adequate clues. Among the possible clues, one in particular interests us here, because it is by itself sufficient. Two simple experiments will make this clear.

Hold a pencil a few inches from your nose and another one at arm's length. Now, look at them alternately with one eye or the other. They will appear some distance apart, and this distance seen through one eye is different from that seen through the other. The difference (hereafter *disparity*) is but a consequence of the way the two objects are projected on the retinas—it's just how things work out geometrically because our eyes are in different places in our head and view the world from different angles. But that disparity, and this is the crux, can serve as a clue to the relative depth of the objects.

Indeed, the way Evolution engineered things, it is the perfect clue: she endowed our eyes with a finely honed feedback system controlling their movement, so that whatever we fix on, its image will fall on the center of the retinas (see figure 14.8); any point behind or in front will project at some distance from the retina centers, and the distance and direction of those projections are uniquely determined by the positions of the points relative to the fixation point.*

It is this full determinism that makes disparity a peerless clue. You can get a feel for its information power by taking a pair of photographs of a scene from slightly different angles and viewing them through an old-fashioned stereoscope. This is a simple optical device, basically just a set of adjustable mirrors allowing you to present the two pictures separately to your two eyes. You needn't manipulate the disparity between the pictures; disparity will manipulate you. All you need to do is adjust

---

*Other image parameters, like perspective and relative motion of near and far objects (parallax), can also offer depth clues, and so can edges and shadows of objects; but the point is, disparity is a sufficient clue. At distances of more than 100 meters, disparity is no longer useful; perspective will then stand in for it.

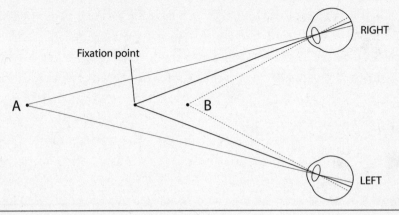

Figure 14.8. Disparity as a Depth Clue. When we gaze at a point (*fixation point*), it projects onto the center of both retinas. The diagram shows the projection of this and a farther (*A*) and a nearer point (*B*).

the mirrors until you see the two flat images fuse, and depth will literally spring to your eyes.

That experiment isn't completely tight. One may argue that other image features, like perspective, shadows, and edges, provided clues. But that argument can be disposed of by using, instead of the photographs, the pictures in figure 14.9. These pictures have no features of this sort. In fact, they contain only random dots, and the dotting is identical in the two pictures, except for a subtle shift to the left within a small square area at the center.*

In short, the two pictures contain no information whatsoever, except for disparity. And that is all the information our brain needs: fix on the pictures with both eyes until you see them fuse, and depth will bob up at the center.

Now, returning to our evolutionary question, a plausible reason presents itself for the last-minute brain reorganization: disparity. Indeed, if we trace the information flow in the visual system from its sources outside (figure 14.7), we see that the rearrangement of sensory lines brought together information on the cortex that originates at the same outside

---

*The dotting was generated by a computer program devised by the psychophysicist Béla Julesž.

Figure 14.9. A Random-dot Stereogram. The two pictures are identical, except for a slight shift to the left within a square area at the center, creating an artificial disparity. Viewed with one eye, neither picture shows a texture, but if you use both eyes and allow the images to become fused, you see a central square floating above the background. It may take a while until you get a fusion, but it goes much faster upon subsequent trials. (From Julesz 1971.)

locales—precisely the kind of information necessary for computing disparity. Thanks to the rearrangement, that information now ends up in the same cortical locale, whereas formerly it ended up far apart in different brain hemispheres.

Thus the disparity information can now be directly processed by neighboring cortical cells, obviating the necessity for a long inter-hemispheric route. This makes for a faster depth computation—an advantage that must have weighed heavily in evolution, as the brain-cell mass grew and the length of the inter-hemispheric route became a limiting factor in the speed of the computation.

And that is the advantage that all the creatures with forward eyes enjoy . . . except for the hapless Cyclops.

Information Processing in the Brain from the Bottom Up

The processing of disparity takes place at the first way stations of the sensory-information stream in the cortex (V_1, V_2, and V_3; figure 14.4, bottom). That piece of knowledge comes from experiments in which cells in the primary visual cortex of monkeys—cells receiving input from both eyes—were probed with microelectrodes while pairs of pictures of varying disparity were presented to the animals' eyes. The cells fired bursts of electrical pulses in response to particular disparity values, and the number of pulses fell off sharply with either higher or lower values—the cells were attuned to a particular disparity, as it were.

We are getting here a glimpse of the workings of the biological virtual-reality machine foretokened in chapter 12. It's only a small glimpse, but it bodes well for analysis. Despite the obvious intricacy of the information web overall, the information processing involved in basic depth perception is straightforward; it follows the primary sensory information stream from the bottom up, and it is unimodal.*

Such a simple processing mode is difficult to beat in speed. Given the way the neuronal web is structured, a top-down mode has necessarily more stages and longer information lines—every synapse adds about another 5 thousandths of a second in processing time and every centimeter of (myelinated) axon, 1 ten thousandth to 1 thousandth.

Not that there is no top-down processing in the sensory brain. Evolution has found ample uses for it elsewhere (and to the analyst's chagrin), especially in sensory processes giving rise to conscious perception, as we shall see. But here she kept things pure and simple.

Being of One Mind

So far we have treated the brain as if it were a single entity. Anatomically, however, it is more like two organs, a matching pair (the "hemispheres") connected at the base, each hemisphere running its own virtual-reality show and contributing its share to virtual-image computation. The results are integrated somehow, yielding, all in a fleeting

---

*Such simplicity seems to prevail up to V4, V5, and the inferior temporal cortex. Further along the information stream, in the amygdala and orbifrontal cortex, things become more complicated, as the somatosensory, olfactory, and taste streams converge with the visual one there.

instant, a unified whole. So we see a single picture of the world outside, not two.

And here is where the corpus callosum comes in.

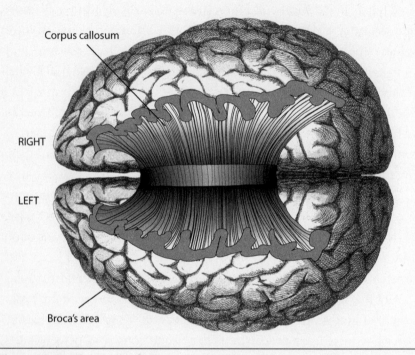

Corpus callosum

RIGHT

LEFT

Broca's area

Figure 14.10. The Corpus Callosum. A top view of the brain with a window cut out, displaying the axons of this massive bundle, which fan out after crossing the midline of the brain.

This brain structure is made of some hundred million axons that link the two hemispheres at the bottom. It is a cable of axons, the largest one in our body. As a whole, this structure had been known for centuries. Indeed, it is difficult to miss; it sticks out by its massiveness and white color (owing to the myelinated axons). Its function became known only some 50 years ago, thanks to the psychologist and neurophysiologist Roger Sperry. He and his colleagues severed the corpus callosum, splitting the brain in two halves, and what they found, it is fair to say, has taught us more about the mind than all 300 years of psychology before.

Sperry and his colleagues began with experiments on cats and monkeys that had been trained to perform well-defined cognitive tasks in response to sensory inputs from either the left or the right side of the body. After splitting the animals' brains in half, the cognitive capacity of each half could thus be tested separately (see figure 14.11). The testing revealed a variety of cognitive deficiencies, depending on the task learned, but they all came down to this: what one half of the brain had learned was not passed on to the other.

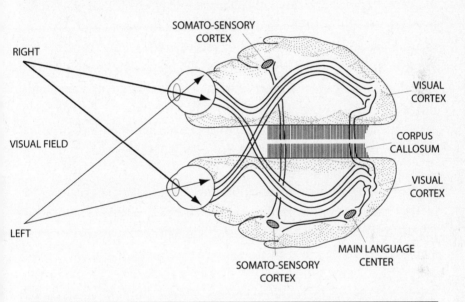

Figure 14.11. Split-brain. Severing the corpus callosum disconnects the language center from sensory input of the left half of the visual field and of the left side of the body (touch). As a result, split-brain subjects cannot name or describe objects in these two sensory fields. The diagram sketches the severed inter-hemispheric information flow between the processing stations in the visual cortex and between the processing stations in the somato-sensory cortex. The layout of the visual information lines and their partial cross-overs are also shown; for the straightforward cross-over layout of the touch information lines, see figure 14.5.

This result in hand, Sperry and his colleagues went on to experiments on human subjects. These were epileptic patients whose corpus callosum had been severed for medical reasons. Such surgery is sometimes performed as a last resort to alleviate intractable seizures. The patients, apart from having fewer seizures, then often report feeling much better also in general. They can go on with their daily lives. Their speech, verbal reasoning and recall, personality, and temperament are all preserved to a surprising degree—one would hardly suspect that there is something wrong with their brains.

But there is. When prodded with one-sided sensory inputs in the manner of the experiments above, the subjects show definite cognitive deficiencies. And they do so most startlingly when, in full possession of their language faculty, they are challenged with sensory inputs coming from the left side of the body or the left half of the visual field. They are then unable to say what they feel or see, woefully failing in even commonplace recognitions. For instance, though they are perfectly able with eyes closed to call a pencil a pencil when it is placed in their right hands, they cannot do so when it is placed in their left hands. Or although they can name and describe objects that fall in the right half of their visual field, they cannot do so for objects in the left half of the field.

The reason for these failings becomes apparent if we follow the corresponding sensory-information lines to the primary cortical processing stations, and from there to the cortical processing stations of linguistic information—the so-called language centers (figures 14.10 and 14.11). There are two such centers in the human brain: one, *Broca's area* in the frontal lobe, we have already seen (chapter 13); the other, *Wernicke's area*, is located in the temporal lobe. But both lie in the left hemisphere. So with the corpus callosum cut, those centers no longer receive input from the sensory processing stations on the right—the information from the left half of the visual field gets only as far as the mute right hemisphere.

Two Minds in One Body?

I mentioned that split-brain patients go on with their lives reasonably well. This will now be all the more surprising in view of the large information deficits they are afflicted with. But the patients learn how to compensate for the lopsided deficits by using additional sensory cues: bilateral cues of touch through a shifting of hands, visual cues through exploratory eye movements, auditory cues, and so on. But those are slapdash remedies—with just a little prodding the cognitive deficiencies surface again.

Or perhaps better said, they burst forth, because they come with strong emotional undercurrents. Consider the following example from an interview with a patient who was successfully operated on, videotaped by the neurologist Victor Mark.

The patient was a young woman who had been given a sheet of paper with the words *Yes* and *No* printed on top and instructed to use it to signal her answers. When she was asked, "Is your left hand numb," her left hand pointed to *No*, and the right hand to *Yes*. She became visibly annoyed with the lack of unanimity in the results and repeatedly tried to indicate the "correct" answer, alas with the same results.

Elsewhere in the interview, she was given tallying problems and instructed to signal the results with the fingers of her hands. When asked how many seizures she had recently experienced, her right hand lifted three fingers, her left hand, one. As before, she appeared astonished herself by her conflicting responses. Upon repeat performances she became visibly agitated, her frustration and emotional upheaval punctuated with tears.

The subject's two brain halves evidently held different opinions— one half didn't know what the other one knew!

An Old Pathway Between Brain Hemispheres
for Information Producing Emotion

Not often, but every now and then, nature allows us a peek into the arcane world of the mind. Here she shows us an astonishing autonomy of the brain hemispheres, and it is not only an autonomy of immediate perception but also including memory—each hemisphere had its private memories, whole chains of memory, that had become inaccessible to recall by the other hemisphere after corpus-callosum transection.

However, we should be aware that this is an autonomy of sensory-information processing and, up to a point, of virtual-image computing, not necessarily of consciousness. I wish to stress this point, lest one read too much into the split-brain results. No few workers have fallen into that trap and concluded that the autonomy includes consciousness—or to put it bluntly, that consciousness can exist in half a brain. The split-brain experiments provide no grounds for that conclusion. If anything, they may support the opposite conclusion.

Indeed, we need look no further than a variant of the foregoing experiments by Sperry and his colleagues, in which the subjects were tested with information generating emotional reactions. In one kind of experiment the subjects were shown embarrassing photos. When the images fell in the left half of the visual field, the subjects were unable to see them, yet they would display the appropriate emotional reaction (blushing, etc.).

In another kind of experiment, subjects were tested with offensive odors that were presented either to their right or left nostril. Here the subjects were able to name the odors smelled through the left nostril, but not those smelled through the right one (unlike the information lines of vision, those of smell don't cross over; figure 7.1), while the emotional reactions ensued regardless of what nostril provided the sensory input.

So, information giving rise to what is perhaps our most basic conscious state, emotion, still gets from one hemisphere to the other when the corpus callosum is cut. There must be another neuronal pathway further down, and the indications are that it goes through the brain stem—a passage below deck, so to speak. It goes back to a time in brain history when the corpus callosum still wasn't there—nor, for that matter, was the cortex. The centers for sensory and motor functions were then all in the brain stem. The corpus-callosum axon cable came into existence only with the advent of the cortex and grew hand in hand with it.

The Virtues of Parallel Computation

We now can begin to look back at the brain's sensory operations with some perspective. The primary operations are bottom-up information processings by neuron clusters forming networks. In the grey mantle of the brain, the networks typically have multiple inputs where con-generic information is processed in parallel. This enables the system to perform many computations at a time—a very different operational mode from that of our technological computers, in which information is processed sequentially, one piece at a time.

Parallel computation has its advantages. The most obvious is speed. That on its own must have tipped the scales of Evolution toward selection. But there is another, more basic advantage: information economy. Neuronal parallel computation is inherently cooperative, as the network units share the workload and information cost of computation.

From our perch high on the evolutionary ladder, it is hard to see when exactly the parallel mode came into use. But it must have been early, for we find it in neuron systems everywhere, from worms to humans. The systems vary in cell number from a hundred to a trillion—an immense range, reflecting a growth unmatched by any other type of cell. Throughout that growth, their subnetworks got molded and re-molded again and again for different uses, but whatever the mold, at bottom those networks remained what they always were: tools for parallel computation.

As the saying goes, the more things change, the more they stay the same.

The alternative, sequential computation, really never stood much of a chance. The dies for parallel computation were cast in Evolution's workshop the moment the lady bet her bottom dollar on neurons—a rather sluggish material as logic switches go (a million times slower than silicon chips). So it was either parallel computation or a world of sloths.

Evolution certainly has come a long way with that choice, to judge by the speed of our senses, especially vision. The eyes are our paragons of fastness—we use expressions like "in the blink of an eye," "in a twinkling," or "quick as a wink" without giving them a moment's thought. But how fast is fast? How long does it actually take for us to see something? We need hard numbers here to get onto the cognition process. Those numbers are not easy to come by. Most speed estimates of cognition require a leap of faith, as they are based on measurements of times of body reactions including times of muscle movement. However, one type of estimate comes close to the mark. It is based on a measurement taken on an animal from the lower zoological ranks, and it speaks volumes.

This measurement forms part of the elegant work of the biophysicist Werner Reichardt and his colleagues on the cybernetics of flight of the house fly. The fly is small brained, but its visual system is well developed. The system steers a flight apparatus by means of which the beastie does a number of charming things, like landing on people's noses or escaping a swatter. The apparatus drives a set of muscles attached to the wings, which control the lift and forward thrust of the fly, and the torque that is generated by the asymmetries in thrust from the left and right wings turns it sidewise.

The onset of this torque will serve as our index of visual reaction. The torque was precisely measured by the Reichardt group under varying visual-input conditions, and its onset, that is, the time elapsed between application of the visual stimulus and the earliest detectable torque, provides a reasonably close measure of how long it takes the brain of the fly to process the visual information. The times measured averaged 21 thousandths of a second.

Somewhat longer times have been measured in experiments on human subjects, where the change in electrical brain activity elicited by a standardized visual input served as the index. The subjects were asked to recognize various animal categories in photographs projected on a screen, while the electrical brain activity was led off with metal plates placed at various locations on the scalps of the subjects. The times from visual stimulus to detectable electrical change averaged about 150 thousandths of a second. Of these an estimated minimum of 35 thousandths went into the information transmission from the retina to the cortex (V1), leaving about 100 thousandths of a second for information processing.

The message from the two experiments is clear enough: the observed speeds cannot be accounted for by sequential computation. In both cases, especially the one with human subjects, the visual-information input was complex, requiring multiple operations of processing and computation that could not have been achieved serially within the mentioned time limits—each sequential step, each synapse alone, would add 5–10 thousandths of a second.

So, if the slow logic-switch rate of neurons had worried you, it is more than made up for by their parallelism in computation. We are so used to the sequential mode of our home computers that we tend to be overly concerned with such rates. But switch rates aren't all that important when we deal with computers structured for parallel computation. Standard engineering handbooks won't tell us much in this regard; they are largely geared to sequential computation. Engineers here would do well to take a leaf from Evolution's book—the brain of a fly using parallel computation will do a hundred billion operations per second while merely resting.

It takes a while to be at home with parallel computation. But its advantages are unmistakable, and we'll see more of them further on as we deal with quantum computers. Well, by the time we get to the end of the book, you may want to trade in your laptop computer!

CHAPTER FIFTEEN

Information Transforms in the Cortex and the Genesis of Meaning

With the general principles of information processing in the brain behind us, we are in a position to delve a little deeper into the mysteries of its grey and explore how that neuron web extracts meaning from the information hodgepodge of the world outside. Let me sketch briefly what is to come in this chapter. We will take a closer look at the higher-order sensory information transforms that the grey renders—the "abstract" versions of external reality—and see how these versions are gradually loaded with meaning as the sensory stream of information ascends to and through the grey.

The Censoring of Sensory Information

Those "abstract" versions of external reality have gone through a heavy-handed censorship whereby raw information without biological significance has been bleeped out. The same may be said for other parts of the sensory cortex. What primarily gets through is biologically meaningful information—smells promising good eating, burblings suggesting a spring, rustlings betraying a predator, tingles rousing sexual desire, and so on. Those are the kinds of things an organism needs to know and the brain wants to know.

As for the censors, these keep a low profile. However, with a little prodding with sharp-pointed electrodes, they will come out: they are deployed strategically along the sensory streams, at the output side of the synaptic way stations. We have met their likes earlier on, the neurons with branched dendrites that receive both excitatory and inhibitory inputs (figure 12.2). The dendrite membrane integrates these inputs, and the result of these computations determines the neurons' firing rate of electrical pulses; the excitatory inputs tend to raise the rate and the inhibitory inputs, to lower it. This statistical game decides what information gets through a way station and what does not.

This makes for a vigorous censorship, as that statistical game is played again and again along the way to the cortex. And it doesn't end there. In fact, it is at its best inside the grey. There, the way stations are manned by neurons with particularly long dendrites receiving hundreds or thousands of inputs (figure 14.3). These neurons grind away at the integrations of the inputs to the limit of saturation, playing the game with utmost finesse, though the game never loses its unabashed utilitarian character. Evolution has seen to that from the start and woven the banausic biases right into the neuron web.

All sensory systems participate in the game. The eyes play it somewhat differently than the ears, and the nose differently than the sensory skin, but in principle it is the same statistical game. The end result is certainly the same: a streamlining of the information flow as it ascends to the cortex. Overall this may give the impression of information filtering. Indeed, close to the sensory periphery there is genuine filtering, notably in our sense of touch, where the filters are purely mechanical and passive, as we shall see. But as soon as we get to the synaptic way stations, any filter analogy wears thin, because the streamlining is no longer passive, but rather the result of computations.

Such computations have a stiff thermodynamic price. This imposes huge information demands on the neuron web—and under the stern Second Law, these demands can be neither postponed nor fobbed off. There is thus a considerable influx of extrinsic information to that end, though it all runs automatically and smoothly, thanks to whole armies of molecular Maxwell demons beavering away in the dendrite membranes and silently paying the thermodynamics price for us (see chapter 12).

What the Eyes Tell the Brain

Thus, the streamlining of sensory information in the brain is an active process, which is why I called it censorship, and I know of none more strict. Indeed, it is the very model of censoring: the bowdlerizing, the overseeing, the special veil of secrecy, even the laying down of the rules.

That basic bit of knowledge had escaped the philosophers of the mind—and there was a time when they had their place in the sun. They had no way of knowing about the censorship, so generations of them, "idealists" and "realists," had it wrong. Two hundred years of exegesis of the mind need never have occurred, and a lot of clever men could have spent more time in the fresh air. They simply had no clue that our senses don't tell it like it is.

In fact, no one before the 1950s had. That knowledge had to wait until appropriate tools for tapping the sensory information flow in the brain were on hand, and it took a master, the neurophysiologist Stephen Kuffler, to do the tapping. Kuffler had blazed the trail of single-cell electrical probing at a number of strategic places of the nervous system, and in the early 1950s he set his sights on the ganglion cells of the eye, eavesdropping on the visual information stream in the cat at this early way station. The ganglion cells are relatively large and make good targets for microelectrodes. They are located just under the white of the eye, and Kuffler pierced the white with his electrodes from the outside and led off the cells' digital signals while shining light into the eye.

I still remember these experiments as if it were yesterday. It was in the crowded basement of Johns Hopkins University's Wilmer Institute. Eight of us were wedged together in there. My pint-sized lab was next to Kuffler's, separated only by a pâpier-maché wall, and I could hear the tat-tat-tat of the loudspeakers connected to his recording equipment. So when the pitch changed, I knew at once that something was afoot and would run next door.

The ganglion cell was sending a message to the brain. But contrary to everybody's expectations, this didn't happen when the whole of the eye was illuminated. In fact, the worst thing one could possibly do was to flood it with light. The cell then just kept firing at its resting rate (a low-pitched tat-tat-tat, about 5 to 10 pulses per second). Only when light was confined to small spots on the retina, encompassing but a subpopulation of the photoreceptors connected to the cell, did it fire at a higher rate. The cell would tell the brain only what *it* wanted it to know.

Kuffler was an experimenter without equal, and in short order found out what that was: contrasts of light and shadows—the chiaroscuros of the world outside. It's what a painter seeks out, and both cell and painter indulge in little heightening touches.

The next steps—and they were giant steps—were taken by the neurophysiologists David Hubel and Torsten Wiesel. They went on to the next way station, the geniculate nucleus, and then to the cortex. Hubel and Wiesel used microelectrodes made of hard tungsten wire, which they positioned, with the aid of stereotactic equipment, stably onto the target cells and listened in on their signaling. They managed to do so for long periods—sometimes nine hours.

The cells in the geniculate nucleus turned out to behave much like the retina ganglion cells; their responsiveness to light contrasts (and sensory fields) were similar. The big surprise came in the cortex. There, some of the cells did something so novel, so utterly original, that I doubt anyone would have thought it possible before: they recognized the orientations of things outside—they were capable of abstracting a particular attribute!

Not that these cells were no longer interested in the light and shadows. They were as interested as the cells at the way stations before them were. But they now had a distinct preference for certain spatial orientations. Hubel and Wiesel discovered this when they shined light through a narrow slit that cast a sharp, though faint, shadow on the cat's retina. The cells would respond then only if the slit had the right tilt. Change that angle by just a few degrees, and the cells would cease to respond (see figure 15.1).

Indeed, these cells are quite finicky in their angle preferences; they will discriminate a difference of about three minutes on a clock. In the visual cortex of monkeys, which more resembles our own, such cells come in two types: one responds to stimulus position in addition to stimulus orientation (the slit has to be both critically tilted and critically

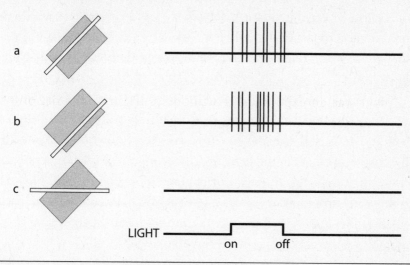

Figure 15.1. A Cortical Cell That Abstracts a Stimulus Attribute: Orientation. A light line on a dark background produced by a narrow slit of light elicits a burst of electrical signals (*right*) at a particular orientation (*a, b*), but not at another orientation (*c*). Each vertical line is a digital pulse on a compressed time scale. The diagrams on the left give the orientation of the slit and its position within the cell's sensory field (*gray rectangle*). This is the response of a "complex" cell; a "simple" cell (not shown) might not signal optimally, if at all, in situation *b*. (After Hubel 1995.)

placed in the cell's sensory field in order to elicit a response), and the other type responds to a specific tilt regardless of stimulus position (figure 15.1). The latter is the predominant type in the primate primary visual cortex—Hubel and Wiesel called them "complex cells" and the others "simple cells." The complex cells achieve a high degree of abstraction, the highest degree attained up to this point of the sensory stream. But we haven't seen the end of it—wait until we meet the other members of that amazing cortical-cell fraternity.

Coding for the Vertical, the Horizontal, and the Oblique

These experiments reached deep into our psyche. They held the mirror up to the mechanisms whereby we acquire knowledge about the outside world, and they go to the ground floor of our sensory abstractions. Seeing is not the only way to acquire knowledge of the world, of

course, but by far the most efficient one—and in a world of sea hori-
zons, ragged rocks, branched trees, and upright man-made objects,
what could be more essential than abstracting the attributes of spatial
orientation!

Kant would have been pleased, I think, by the turn his *Ding an sich*
took at the hands of those who, armed with microelectrodes, bearded
the lion in his den. Cells in the gray mantle of our brain do the abstracting, and chief among them are the complex cells. These represent
all orientations—the vertical, horizontal, and every possible oblique.
Large-scale samplings of the complex cells' digital electrical responses,
which Hubel and Wiesel took in the monkey cortex, showed that all
orientations are about equally represented by these cells.

The marks for this function are neatly laid out: the cells are systematically ordered, forming vertical columns. That by itself might not have
raised many eyebrows outside the anatomists, but the orderliness is un-
mistakably related to the abstracting function: the cells in any given col-
umn code for the same spatial orientation; they are stacked upon each
other in the columns, as the microelectrode probings showed, according
to their orientation preferences—the brain version of computer files.

As for the mechanism of abstraction, this still hangs in the balance.
But at least one can now make models, and model-making has always
been half the fun in science. An elegant model, a simple feed-forward
scheme that takes full advantage of the bottom-up processing mode of
information, has been advanced by Hubel and Wiesel (figure 15.2): the
complex cells here collect the outputs from simple cells that have the
same orientation selectivity but have sensory fields that are slightly offset with respect to each other.

The Neuron Pecking Order

There is streamlining of sensory information along the route from
sense organs to cerebral cortex—a note we already sounded—but
nowhere is the streamlining as pronounced as at the level of the complex cells. The disproportion between the amounts of information going in and coming out of these cells is indeed large. Inasmuch as one

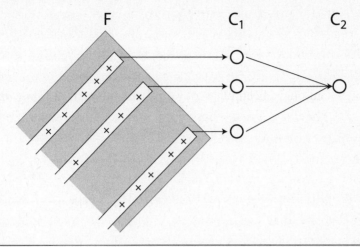

Figure 15.2. A Wiring Scheme for Orientation-selective Cortical Cells. First-order cortical cells (C_1, "simple" cells) collect the outputs from geniculate cells whose sensory fields (F) lie along a line on the retina (see Figure 15.1a, b). Second-order cells (C_2, "complex" cells), in turn, collect the outputs from the first-order ones. The + marks represent centers of partly overlapping radially symmetrical excitatory fields. Each arrow comprehends many axons converging on first- or second-order cells. (After Hubel 1995.)

can glean from isolated cell probings, the simple cells serve the complex cells large dollops, whereas the latter let out but a trickle. However, such blanket comparisons must be taken with a grain of salt. Here we come up against the limitations of Shannon's equation. This comprehends the quantity of information, but says nothing about its quality, its meaning. What a complex cell lets out may be a trickle, but it is loaded with meaning.

I beg the reader's leave to spring this piece of news so suddenly and cavalierly. But I'll make amends; I shall define meaning further on and give an objective measure of it. Here let me just say that what the simple cells ladle out is low-grade, mostly redundant, information. The complex cells weed out such information, and as a result their message acquires more meaning. The weeding already starts with cells at the first way stations of the sensory stream and gets repeated again and again at way stations further on. So if we canvass the neurons along the ascending stream, we find them loaded with more and more meaning. They form a natural pecking order in that regard, with the complex cells on the top.

Such a pecking order also is found in the auditory and somato-sensory systems. There, the flows also get streamlined, and the higher-ups are no less adept at blending low-grade information into a higher grade. The end products are biologically just as meaningful: the pitch of sounds and the direction of touches.

The Grandmother Cell

But how far does the streamlining go, and how far the pecking order? Will the cortical neurons downstream beyond the primary visual cor-tex concentrate even more meaning? Indeed, that is something to be reckoned with if we take the model of figure 15.2 to its logical conclu-sion and project it forward a few cortical stages, to where more and more information lines converge. So the question then becomes, will the information lines eventually converge onto a pontiff in the cell hi-erarchy, a neuron that responds to a whole collection of essential at-tributes of the things we see and recognizes them all at once?

This carries the awesome suggestion of gestalt recognition by a cell—not something easy to swallow, and understandably the possibil-ity wasn't taken seriously at first. However, toward the end of the 1960s enough had transpired about the cells in the grey for neurophysiolo-gists to realize that some of these cells, especially the highly arborized ones, had unprecedented computing capacity. So a few workers ven-tured to check up on those cells, with an eye on gestalt recognition.

By then it had become known that the cells' pecking order goes well beyond the complex cells in V1—four types of higher-order cells had been found along the line from V1 to the inferior temporal cortex (see figure 14.4, bottom). So the idea was born that atop the hierarchy might be cells capable of recognizing a particular face, say, one's grandmother's. Cells like that were jocularly referred to as "grandmother cells." (For a bit of history see box 15.1.)

The idea was widely met with skepticism, if not outright derision. Neuroscientists were of no mind to hear, much less embrace, the notion of gestalt recognition by cortical neurons—the few advocates must have felt like Don Quixote tilting at windmills. But the windmills here were real. Well, that's the way it goes in science. The embryologist Karl Ernst von Baer, a deep thinker of the nineteenth century, once made the melancholy remark (it was near the end of his life) that all new and important ideas in science must pass through three stages: first dismissed as nonsense, then rejected as against religion, and finally acknowledged as true, with the proviso from their initial opponents that they knew it all along.

The plain fact is, gestalt-recognizing cells do exist. The first were tracked down in the late 1960s by the neurophysiologist Charles Gross and his colleagues. These workers probed with microelectrodes the monkey inferior temporal cortex, a locale that receives input from higher-order cells of the primary visual cortex. What they found fitted the notion of "grandmother cells" to a tee: neurons that specifically responded to hands or faces—these cells selectively fired digital electrical pulses when the animals were shown images of these two types of gestalt.

More recently, such cells were also found in humans. Christof Koch, his student Gabriel Kreiman, and the neurosurgeon Itzhak Fried (2000) demonstrated their presence in the human medial temporal cortex. They placed microelectrodes in the amygdala, an almond-shaped region of that part of the cortex, which is known to be involved in our affective states (especially fear), and monitored the electrical activity of cells while the subjects were shown pictures of persons, animals, buildings, and so on.* The results could hardly have been more dramatic: one cell, for instance, would specifically respond to the image of President Bill Clinton; it fired off volleys of electrical pulses when the subject was shown a photograph of Clinton, but not when photographs of other U.S. presidents, famous athletes, or unknown persons were shown (see figure 15.3).

---

*The subjects were neurological patients slated for surgery to alleviate epileptic seizures. The investigators took advantage of the diagnostic monitoring, a week-long period before surgery, in which electrodes are inserted into the patient's brain to locate the focus of seizures.

Box 15.1. The Grandmother Cell

The term originated in a lecture the neurobiologist Jerome Lettvin gave in the fall of 1969 to undergraduate students of MIT. Lettvin, never one to keep his class asleep, illustrated the problem of how neurons represent individual objects, with the following story, which he encored for six years:

In the distant Ural mountains lives my second cousin, Akakhi Akakhievitch, a great if unknown neurosurgeon. Convinced that ideas are contained in specific cells, he had decided to find those concerned with a most primitive and ubiquitous substance—mother. Starting with geese and bears, he gradually progressed over the years to Trotskyites under death sentence in Siberia. And he located some 18,000 neurons clustered in von Seelendonck's area (rarely recognized among the degenerate materialist Sherringtonian exponents in the West) that responded uniquely only to the animal's mother, however displayed, whether animate or stuffed, seen from before or behind, upside down or on a diagonal, or offered by caricature, photograph, or abstraction.

He had put the mass of data together and was preparing his paper, anticipating a Nobel prize, when into his office staggered Portnoy, world-reknowned for his Complaint. (Roth had just published the novel.) On hearing Portnoy's story, he rubbed his hands with delight and led Portnoy to the operating table almost hid under microstereotactic apparatus, assuring the mother-ridden schlep that shortly he would be rid of his problem.

With great precision he ablated every one of the several thousand neurons and waited for Portnoy to recover. We must now conceive the interview in the recovery room.

"Portnoy?"

"Yeah?"

"You remember your mother?"

"Huh?"

(Akakhi Akakhievitch can scarcely restrain himself. Dare he take Portnoy with him to Stockholm?)

"You remember your father?"

"Oh, sure."

"Who was your father married to?"

(Portnoy looks blank.)

"You remember a red dress that walked around the house with slippers under it?"

continues

Box 15.1. The Grandmother Cell *(continued)*

"Certainly."

"So, who wore it?"

(Blank)

"You remember the blintzes you loved to eat every Thursday night?"

"They were wonderful."

"So who cooked them?"

(Blank)

"You remember being screamed at for dallying with the shikses?"

"God, that was awful."

"So, who did the screaming?"

(Blank)

And so it went. Every attribute had by his mother was clear—for no attribute is the substance. To whatever detail Akakhi Akakhievitch pressed, the detail was remembered. Even though the details were compresent, the attributes bundled, somehow or another Portnoy looked blank when asked if they suggested his mother. That there was a simple person to whom all the attributes belonged was freely conceded. All the relations had by this person were remembered, including mothering (which is not a mother). It made no difference—Portnoy had no mother. "Mother" he could conceive—it was generic. "My mother" he could not—it was specific.

Akakhi Akakhievitch went into panic. It occurred to him that if there were but one logician or philosopher on the Nobel committee he was doomed. He had abolished the cells for knowing mother as *ding an sich*. He had let the Kantians be his guide and was now trapped.

If he had only gone after "grandmother cells" he would have been safe, simply because grandmothers are notoriously ambiguous and often formless. A patient man, he went back to the geese and bears and "grandmother cells."*

The neurophysiologist Jerzy Konorski, building on Hubel and Wiesel's findings, had put forth the idea of single cortical neurons sensitive to faces, hands, etc., in 1967 in his "Integrative Activity of the Brain." He called them "gnostic neurons" and predicted that they formed fields in the cortex ("gnostic fields"), which, if destroyed, would lead to category-specific loss of recognition.

---

*From a letter by Lettvin to Horace Barlow

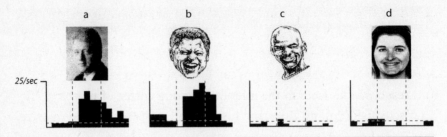

Figure 15.3. Gestalt Recognition by a Cell in the Human Cortex. Digital electrical pulses from a neuron in the amygdala (medial temporal cortex) in response to: (*a*, *b*) a photograph and a line drawing of U.S. President Bill Clinton; (*c*) a famous athlete; (*d*) an unknown person. Each image was shown for one second while the cell's electrical activity was led off with a microelectrode implanted in the brain of the conscious subject. The corresponding responses—the rate of digital signals vs. time—are displayed in the second row. The one-second period of image presentation is marked off by the vertical stippled lines. (After Kreiman 2001 and Kreiman, Fried, and Koch 2002.)

Distributed Coding

On evolutionary grounds, and cell physiology apart, face recognition seems a likely function for a system that abstracts information about gestalt. One can glean lots of useful things from a mien—identity, fear, joy, anger, affection, rejection, approval, derision, suspicion, candor, humor—and all within a fraction of a second.

However, a word of explanation is in order. In speaking of grandmother cells above, I used the singular. I was foxing the reader a bit—I meant it generically. No single cell could ever bring off a feat like gestalt recognition on its own. It wouldn't even get the next cell to fire. Those profusely arborized cortical cells are not particularly trigger-happy. Their long dendrites typically receive information input from many axons—it is not uncommon that thousands of axons, even tens of thousands, impinge on a cell. And because the local electrical effects of these inputs are distributed over a large surface of dendrite membrane, it takes the sum of many inputs to push a cortical cell over the firing threshold (the integration is much like that of the local generator potentials at peripheral sensory dendrites; see figure 4.2).

Rather than operating singly, the grandmother cells more likely pool their outputs and operate in cooperating groups—and it was in that generic sense that I used the term "grandmother cell." The available evidence speaks for such an ensemble action. In extensive samplings of neurons responsive to faces in the inferior temporal cortex of monkeys, the neuroscientist Edmund Rolls and his colleagues (1998) showed that the representation of a particular face is distributed over a set of neurons— for example, a set of 25 neurons would respond to one out of 3,000 different faces with better than even odds.

The pooling of outputs comes naturally to neurons forming networks. It is also highly information-economical, as the participating neurons share the heavy thermodynamic load of cognition—and in a workshop so beholden to economy as Evolution's, that alone may have ensured the adoption of the distributed-representation mode in the brain. That mode amounts to a distributed coding, and at its most economical, the members of the coding ensemble (the set of 25 neurons in the case above) may be attuned in varying degrees to a gestalt.

In a wider sense, distributed coding is part of Evolution's strategy of parallel information processing and parallel computing. The whole brain, the very structure of its neuron web, is geared to that strategy, as we have seen. The parallelizing is most obvious at the sensory periphery. There the information about the world outside is divided up into major channels— the visual channel, auditory channel, somatosensory channel, and so on. Those information channels have long been known. But higher up the parallelizing continues, as the sensory flow is broken up at countless forks. In the stretch from V1 to V4 alone, there are three subchannels, each with its own abstraction function and neuron pecking order: one channel deals with information about shape; another with information about color; and a third with information about movement, location, and spatial organization. And all this is done in the interest of organismic reaction speed.

So when all is said and done, it is the partitioning of sensory information and the parallel processing and computing that allow Evolution's creatures to cope with the challenges of the world outside. And astonishingly, despite long information lines, they do so with time to spare.

Though there is this rare exception:

> *There once was a Dachshund who was so long*
> *He didn't have any notion*
> *How long it took to notify*
> *Its tail of his emotion*
> *And so it happened that while his eyes*
> *Were filled with woe and sadness*
> *His little tail went wagging on*
> *Because of previous gladness.*

The Imprinting of Cortical Cells

Now to a more sober point. How do grandmother cells get encoded? Nobody will think that such cells come wired from birth to know Bill Clinton. So some form of learning is bound to be involved—or in general terms, the encoding is a product of experience.

But what kind of experience? The first thought that comes is an experience that translates to synaptic rearrangement in the neuron web. The connections in the web are not cast in stone. Old connections are broken and new ones form—a perennial game that is ruled by usage. Thus, it stands to reason that the connections of grandmother-cell ensembles, connections conducive to gestalt recognition, form through repeated usage of certain visual pathways.

An incisive series of experiments by the theoretical physicist Tomaso Poggio and the neurophysiologists Nikos Logothetis and John Paul speaks to this point. The investigators trained monkeys to recognize certain objects: paperclips twisted in three dimensions. These objects are as simple as they come. They are but skeletons, almost pure gestalts. Unlike in the experiments on grandmother cells we saw before, the subjects here were not familiar with the objects—paperclips are not exactly fixtures in the monkey world. The animals were trained, by routine reward procedures, on one particular paperclip shape from one particular angle of view. It took on the average four months for them to learn to recognize that target.

It was then that the actual experiments started. The animals were shown the target while the digital electrical signals of individual cortical cells were tapped. A set of cells in the inferior temporal cortex reacted selectively to the target. They produced a vigorous barrage of electrical pulses in response to it, whereas other objects, including different paperclip shapes, elicited no significant signal output (see figure 15.4, top). And when the target was rotated about its vertical axis away from the training angle, the signal output decreased or ceased altogether (see figure 15.4, bottom).

Next, rounding out the experiments, the animals were trained on several angles of view of the same paperclip shape, while the signal output

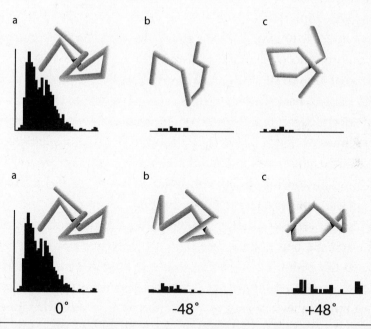

Figure 15.4. The Grooming of a Grandmother Cell. The responses of a cell in the inferior temporal cortex of a macaque monkey trained to recognize a particular paperclip shape (target). The ordinates of the plots are the averaged rates of digital pulses produced by the cell; and the abscissae, the time. *Top row*: (*a*) response to the target and (*b, c*) to other paperclip shapes. *Bottom row*: (*a*) response to the target at the angle of view on which the animal was trained, 0; and (*b, c*) when target is rotated to other angles of view, −48°, +48°. (After Logothetis and Paul 1995.)

of those cells was tested again. They now would respond maximally to all angles of view, in a full 360° circle—and they did so with as few as three training views 120° apart.

From only three examples, the cells had learned how to generalize.

The Begetting of Meaning

These generalizations, as well as the underlying abstractions of gestalt, are associated with large information losses—and I mean losses independent of the red ink mandated by the Second Law. Indeed, a good deal of information goes down the tube all along the sensory stream, especially at the upper echelons of the neuron hierarchy. One might say that's to be expected from the upper classes. But no, not here. Those neurons are denizens of a kingdom where conserving information is noblesse oblige.

If we look more closely, those losses are not wanton. What goes down the tube is information useless to the organism—biologically meaningless information. Aeons of honing in Evolution's workshop have seen to it that all along the sensory stream the wheat gets separated from the chaff. We can't see all the details of the separation process, but this much we can, in bird's-eye view: that meaningless information gets discarded along the way, while meaningful information is conserved, so by the time the stream reaches the upper neuron echelons, it has become quite lean but is heavy with meaning.

Such a bird's-eye view of the stream is hardly sufficient to navigate the shoals of psyche, but if one has for years contemplated a problem without the slightest advance, every bit is welcome. And here we are let in on the secret of the grandmother cells. Their astonishing feat turns out to be as much owed to their losing information as to their extracting it from the stream. By scrapping bits and pieces, they discern the essence of things, the gestalt.

It's like Lewis Carroll's famous Cheshire Cat, who lost his features one by one, until all that was left was the essence, the smile. And it was a smile never to be forgotten.

How to Measure Meaning

And so we arrive at a working definition of meaning, where this precious information quality is generated at higher-order neurons through selective information losses. It is a definition free of the usual anthropocentric aspects; rather than a human privilege, it accords all creatures with such neurons the capacity of generating meaning. It's amusing that this concept should be ecumenical, while what theologians have been plying in this regard for centuries is not. Oh well, so it goes.

However, we are still not quite home free. Missing from the definition is an objective measure. But how does one measure something ethereal like meaning? This question has preoccupied generations of philosophers, but for a scientist there isn't much to take home from their musings in that regard. The physicist Richard Feynman liked to say—and did so with only the slightest twinkle in his eye—that the philosophy of science is about as useful to scientists as ornithology is to birds.

To be fair, classical information theory isn't of much help either, and if you expect an answer from its books, you will be disappointed. Shannon left meaning out of his formulation (see the appendix, equation 3). And he did it with full intent, so as not to mar the beauty of the information concept. His equation tells us how many bits of information are in a message, but nothing about whether those bits convey something shallow or profound.

One cannot help being dissatisfied with a notion that leaves out what to us humans matters most. However appropriate the cold mathematics of that equation may be for engineering, we know in our bones that a thousand bits of nonsense don't equal a thousand bits of sense.

Is there a way out of this quandary? It may seem unlikely that an answer should be forthcoming from physiology—and cell physiology of all places—though it's those dazzling grandmother cells that do throw out the hint. Recall that these cells are richly arborized and perform complex integrations. Such computations entail a good deal of work—no one, not even a pontifical neuron, gets a free ride. So the thought lies near that meaning may be measured in terms of the work going into these computations.

Fortunately we needn't start from scratch here. We may lean on computation theory. This modern branch of mathematics offers a measure that comes close to what we are angling for: a yardstick for structural complexity—and under the information looking glass, complexity and meaning are birds of a feather.

Logical Depth

That yardstick is the brainchild of the mathematician Charles Bennett. He named it *logical depth*, and it gauges the complexity of a structure in terms of the computational work needed to generate it. Computers can generate virtual structures, and the more complex the structures are, the longer it takes to generate them. The computation time thus provides a measure of complexity, and Bennett defines logical depth in terms of this time: *the time required by a standard Turing machine to generate a given structure* (see the appendix, equation 5).*

The measure may be applied to any structure, whether it be a fullfledged object, a minimalist gestalt, a message, or an information state encoded in a set of neurons. A structure is said to be "logically deep" when it took significant time (meaning significant computational work) to generate it.

So much for logical depth from the standpoint of the originator of the structure or the sender of the message. This is the mathematician's cynosure—he tends to view logical depth largely from the standpoint of the sender of the message. But what of the receiver? The receiver's standpoint is really what interests us most, and viewed from that angle—one might call it Evolution's angle—logical depth is essentially an information-economizing device. It saves the receiver computational

---

*The seeds of the logical-depth concept were planted by the mathematicians Andrei Kolmogorov (1965) and Gregory Chaitin (1977) in algorithmic information theory, an offshoot of classical information theory. In their work, structural complexity was measured by the length of the minimal computer program generating a structure. Bennett's concept goes further. It takes fair account of all programs capable of producing the structure as output, not only the minimal one. It defines logical depth as the minimal computer time required to generate a structure from an algorithmically random input.

work; by dint of the logical depth of a message, the receiver gets the hang of it without having to go through the trouble of the computations again. And as far as the receiver is concerned, that is all that matters. She doesn't need or want to know how much effort went into the generation of the message—the message might as well have been produced by the proverbial horde of monkeys hammering away at typewriters!

In short, *logical depth, from the receiver's angle, defines the value of a message as the amount of computational work done by the sender, which the receiver is spared from having to repeat.* And that, we may imagine, is how Evolution saw it when she made her choice. Ever the skinflint, she had set her sights on saving the replacement cost of the message.

We hear the echoes of this parsimony every day in our language. We trim our messages of unnecessary baggage and squeeze meaning into relatively few words. The recipient need not spend undue time or effort; she will get the message, as long as it has enough logical depth.

However, it may strike you as odd that the instilling of logical depth into these messages, which is bound to involve considerable computational effort, takes little conscious effort. The scanting of information, the neuronal computations, and whatever else goes into the production of logical depth comes so easily to us. To state the truth, we are spoiled. Evolution has bedded us in the luxury of a neuron web that is ready-programmed to run those operations quasi-automatically.

However, the preprogramming will go only so far. It is adequate for quotidian messages, but doesn't cover our more deliberate messages. A scientific paper, an engineering report, or a brief to the High Court is wont to brim with logical depth, but it may take burning the midnight oil to produce that depth. To make things easier, we resort to several time-honored props, and chief among them are the systematic arrangements of data in columns, the tables. But a good table is more than just a systematic arrangement; it has logical depth. Take the tide tables, for example. Every number in them is a compacted message, the result of computations based on Newton's equations. Yet these messages—the heights of the coastal waters for every hour of the day, for a whole year—get across to us at once without having to spend computational effort again, thanks to their logical depth.

Or consider the ephemerides, the tables astronomers use to look up the position of the moon, the planets, and the stars. These tables, too, are based on Newton's equation and have even more logical depth. They are the result of lengthy computations that the user of the tables doesn't have to do again. Owing to their logical depth, he gets the message at once. And speaking of Newton's equations, these themselves brim with logical depth. They (and the initial conditions) contain all the information the ephemerides and the tide tables have, and then some.

Indeed, here we come close to the farthest bounds of logical depth, the peaks of the human intellect. And that's setting the bar a bit high. But if you will settle for garden-variety logical depth, and if it is ready-programming you want, then what the neurons in Broca's area of the cortex have to offer is not bad. When they pull out the stops in their verbal acrobatics, they draw up what comes in from the information hodgepodge of the world outside and from internal brain stores and disgorge the scanted, meaningful information in one stupendous dose.

Lowbrow Meaning

At this juncture an old friend of mine, the Pacinian corpuscle, knocks at the door, to be let into the plot. This ubiquitous little sense organ has its own story about meaning to tell. Till now, our focus has been mainly on meaning arising in the cerebral cortex, and for good reasons; what those neurons up there deliver in logical depth casts everything else into the shade. But meaning is also generated along the way to the cortex—indeed, all along the way—and the Pacinian corpuscle is among the first neuronal elements to contribute its share. And compared to the cortical higher-ups, its contribution is biologically no less important—it is actually more basic, I dare say.

So, if you still can lever up an eyelid, we'll take a closer look at what this tiny sense organ tells the brain.

No question, this sensor has a large presence. It is found in the skin, in muscles, in joints, and in the web the intestines are bedded in (mesentery), everywhere supplying information about local mechanical events. The anatomist Filippo Pacini, while still a medical student, described it in 1831, correctly positing its sensory function. He presented his work to the Academy of Florence, then one of the leading scientific bodies in Italy. Alas, as such things happen, the work was rejected— those corpuscles surely must be parasites, so went the comment of the august body.

To tell the truth, with its large, oval capsule visible to the naked eye, it might easily be taken for a parasite (see figure 15.5). And in a sense the capsule *is* a parasite: it preys on information. But it's information we don't need, as we shall see.

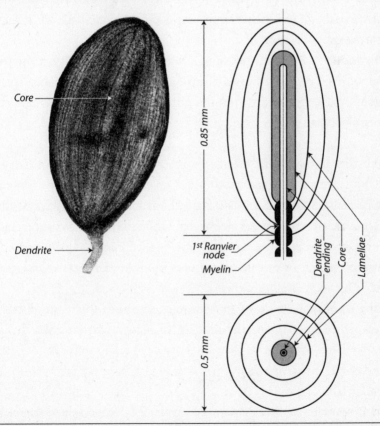

Figure 15.5. Pacinian Corpuscle. *Left:* Phase-contrast micrograph of the live isolated sense organ. The nonmyelinated dendrite ending, the sensor, is seen inside the multilayered capsule together with a stump of the myelinated dendrite dangling outside. (From Loewenstein and Rathkamp 1958.) *Right:* Diagram of a longitudinal and transverse section. The lamellae closest to the dendrite ending are densely packed, forming the "core." (From Loewenstein and Skalak 1966.)

The organelle is very much a sensory one, precisely as the young Pacini had assumed. The hub of the sensory operation lies smack in the middle of the organelle. It is a nerve ending about half a millimeter long—the unmyelinated terminal of a dendrite of a neuron located in a ganglion alongside the spinal cord (figure 14.5). This unmyelinated nerve ending is the receptor proper; it transduces mechanical into electrical energy. Its cell membrane contains, over much of the ending's length, specialized ion channels that open up when the membrane gets stretched, thus transducing membrane strain into an electrical current. This is the *generator current* we are already acquainted with (figure 4.2). And the name is apt: this current sparks off the digital electrical signals that travel to the brain.*

But perhaps I should say signal, rather than signals, for more often than not the receptor fires just one digital pulse. The Pacinian corpuscle is the ultimate minimalist, and in matters of meaning, as we have seen, less can be more.

It's worth noting, as an aside, that the organelle may be readily dissected out of the transparent mesentery membrane of animals and kept alive in a suitable salt solution for several hours. In vitro, its generator current and digital signals can be conveniently led off the dendrite with the aid of small metal electrodes, while the capsule is subjected to calibrated compressions in the physiological range (ten thousandths of a millimeter). In our laboratory at Columbia University, we used a piezo-electric crystal to that end.

The minimalist bent of this sensory organelle thus shows itself at once: the organelle maintains electrical silence in mechanical status quo

---

*The digital signals arise at the first Ranvier node of the dendrite or close to where the myelin starts (figure 15.5, right), membrane regions with sodium- and potassium-channel populations of the sort described in chapter 3. The mechano-sensitive channels that populate the major part of the membrane of the unmyelinated nerve ending are less ion-selective; they will admit several species of ions in their open state. They are sensitive to distortional strain, but not to hydrostatic stress. Presumably the channels are mechanically coupled to the ending's membrane, and the strain gets directly transmitted to them. In a number of invertebrate mechano-receptors, where the channel protein could be cloned, the protein is tethered to elastic components of the cytoskeleton or the extracellular matrix.

and will break it only when the status quo is broken. Typically, it will fire one digital pulse upon compression and another upon decompression, but during the time the compression is maintained, and regardless of how long, it gives no digital response. It is easy to see why: the generator current decays during compression; it falls below the threshold for sparking digital pulses within about a thousandth of a second of the onset of compression and doesn't rise again unless a compression (or decompression) occurs anew (see figure 15.6).

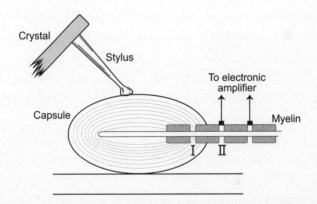

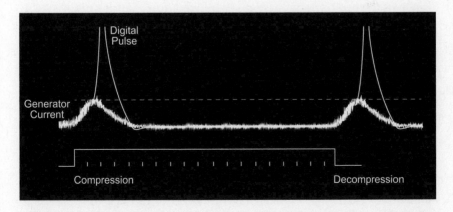

Figure 15.6. *Top:* A Pacinian corpuscle is compressed with a glass stylus driven by a piezo-electric crystal. *Bottom:* time course of the compression and that of the electrical responses—generator current and digital pulse—as recorded from the dendrite outside the capsule.

The question then is, why does the generator current fall during the steady period of compression? Here is where the capsule comes in. It acts as a filter of slow mechanical events. Indeed, its structure is splendidly designed for that. It consists of many thin, elastic tissue lamellae concentrically arranged around the nerve ending, and the narrow spaces between the lamellae are fluid-filled. That structure will efficiently transmit viscous pressure to the ending—a pressure developing during the compression as the fluid is forced to move through the interlamellar spaces (see the appendix, item 7, for a mechanical analog). The viscous pressure is transmitted with acoustical speed through the capsule. Thus the nerve ending's membrane almost instantaneously experiences a strain at the onset of the compression—a strain that the ending's mechano-sensitive ion channels will transduce into a generator current. But by the time the compression has reached steady state, viscous pressure no longer develops, and there is nothing for the mechano-sensitive channels to be transduced into. The only forces then prevailing in the capsule are the elastic ones generated by the lamellae. But these forces are poorly transmitted by the capsule structure—so poorly that, within two milliseconds of the dynamic compression phase, the pressure at the center of the capsule is zero.*

Clearly the capsule calls the shots here. And not only here, at the entrance of the brain's somato-sensory information channel, but by virtue of its strategic location, also over the whole channel; the capsule single-handedly decides what of the outside world the sensory nerve ending gets to see and hence what the brain gets to see. It does so consistently and, in what concerns the cognitive function of our brain, dependably. Its filter mechanism—purely mechanical and passive, though regular as

---

*During the release of the compression—the "off-phase" of the stimulus—a viscous pressure develops again. The capsule then snaps back to its original shape by action of the elastic forces generated by the distended lamellae. The pressure distribution in the capsule is then rotated by 90 degrees, reaching a peak at the capsule center, which is comparable to that in the dynamic phase of compression. Thus, the dendrite ending experiences a strain anew, giving rise to a new generator current and digital pulse—the "off-responses" of this sensory organelle.

clockwork—purges biologically meaningless environmental information, while passing meaningful information. It purges information about environmental status quo and passes information about environmental transients, the sort of information that is useful to the organism.

In short, the capsule substantially contributes to the genesis of meaning in the brain.

But why did Evolution go to all the trouble of making the capsule (and it was late in her neuronal enterprise that she did), rather than just relying on the old mechanism of sensory demon-tandem uncoupling for genesis of meaning (see figure 3.6)? The uncoupling mechanism certainly isn't lacking at the sensory ending here. If anything, it is more highly strung, ready to stop the ending's digital-pulse production the moment the generator current drops below a physiological minimum. So why have two mechanisms when one might suffice?

This takes us into teleological territory, where one must tread carefully, lest one fall into the trap of bringing one's own feverish simplifications to the table. However, here the clues that Evolution offers us are strong enough, and the analysis of the pertinent mechanisms can be carried out with enough detail, to catch on to her designs.

One clue is provided by the kind of environments this sensory organelle inhabits—the skin, mesentery, joints, and muscles. These are environments in which things pulsate, throb, and vibrate. So we ought to look at how mechanical information gets transmitted to the sensory ending under those conditions. The vibrational condition may be readily mimicked in vitro—it's a simple matter of applying sinusoidal mechanical pulses to the organelle with the aid of a piezo-electric crystal (see figure 15.6, top)—and the pressure field in the capsule can be determined as a function of vibrational frequency. The question then is, how does the pressure at the capsule center vary with frequency—or to come straight to the teleological point, at what frequency is that pressure highest?

That maximum works out to be at a frequency of about 200 per second—squarely in the range of the vibrations prevailing in our skin under physiological conditions of touch. The capsule thus serves to attune the sensory organelle to the normal environmental vibration frequency.

So we can see it now—that tuning gives away Evolution's game. It was meaning, biologically meaningful mechanical information, that she was after when she engineered that stupendous little capsule.

CHAPTER SIXTEEN

The Conscious Experience

In his *Ars Amatoria*, the *Art of Love*, Ovid, after spending hundreds of lines on the subject of how best to win a lady, yields to despair and exclaims, "I was about to finish, but different ladies have different kinds of minds." The poet playfully uses the word *pectora*, which in Latin means both breasts and minds. The Romans, like many people in antiquity, believed the mind was in the chest—a notion that would live on through the Middle Ages and Renaissance and, in turns of phrases about the heart, even through our time. It took the levelheaded work of neurophysiologists and neuropathologists in the nineteenth century to give the quietus to that stubborn belief.

What we take for granted today, that mind resides in the brain, received its footing really not very long ago—it evidently takes more than a brick to the head to be convincing in such matters.

But that's about where certainty ends. Regarding the nature of the very foundation of mind, consciousness, we know as much as the Romans did: nothing. We don't even have a valid paradigm. Painful as it is, I say this at the start, so as not to raise false hopes (there has been enough of that in recent years). But it would be unfair to lay all of the blame at the door of the biologists. The gap of knowledge, as I see it, falls as much into the field of physics as it does in that of biology. Consciousness provides a natural meeting ground for biologists and physicists—it is the ultimate frontier.

In this and the following chapters, we'll survey some areas of overlap between the two fields, areas with complementary bits of knowledge that might help bridge the gap.

The Consciousness Polychrome

Since the early 1960s searches have been going on at NASA and various university observatories for signs of extraterrestrial intelligent life—the SETI projects. These are concerted efforts to scan the sky for electromagnetic emissions with patterns standing out of the cosmic static, emissions with high signal-to-noise ratios, which are not attributable to ordinary physics phenomena we know about. If these efforts are successful, and it is a big if, we might then beam out messages of our own. One such message would surely be about life—what and where life is on our far, far off planet. And that is something we could handle; we may perhaps manage even a passable definition. Alas, on the subject of the nature of consciousness, life's finest offshoot—or some would say, its reason for being—the message would be blank. We would be unable to provide anything even remotely resembling a definition. At best, we might give a description of subjective experiences and their neurophysiological and neuropathological correlates, and not without hedging it around with all sorts of qualifying clauses.

Yet we all have an intuitive feeling of what consciousness is. It is a feeling of awareness of the world outside and of ourselves, a feeling with an endless number of nuances—the shades of green of a meadow, the smell of honeysuckle, the sounds of rushing wind. It is a feeling involving the remembrance of things—a beloved face, the touch of a vanished hand, the sound of a voice that is still. And it is a feeling involving the awareness of the passing of time, our joys and sorrows, our wonderings and will—the whole polychrome of the I.

What physics mechanism could possibly give rise to that?!

Consciousness, Sensory Information Processing, and Computing

Let's put physics aside for a moment and look at the development of consciousness. Although from our lookout high up the zoological ladder we cannot see at what rung consciousness first appears, this much we do see: it is the culmination of Evolution's information enterprise—

phylogenetically and ontogenetically. Piece by piece, the evidence in the foregoing chapters stacked up that consciousness is the culmination of the action of highly evolved cortical neurons—the culmination of their information processing and computing.

I use the word *culmination* because it does not have the mechanistic undertones words like *result* or *product* have. All it implies is that information processing and consciousness are sequential events—and that is all we know. This is good to bear in mind at every turn, lest we are led down the primrose path that has tricked quite a few computer enthusiasts into thinking that the two events are identical or that the information processing itself explains the sequel. We live in a technological age of high expectations from computers—and justifiably so. The computers that engineers have been parading have huge information-processing capacities. X3D Fritz, for example, a chess computer, is able to evaluate three million positions per second, without a slip or error, and it has played the world chess champion Garry Kasparov to a draw. But offhand, I cannot think of anyone who would ask it if it is aware of things, if it has tearful feelings, joys, or regrets, if it is aware of the passing of time.

No, there are unfathomed deeps between digital information processing and consciousness, and no amount of knowledge about the processing itself could fill those.

The Conscious Experience of the Passing of Time, and Memory

The situation is not as dire on the Darwinistic-physics front. There our previous forays have already secured us a slight foothold. They brought in the reason for being of the sensory information processing, its evolutionary role: *to provide the organism with a biologically meaningful representation of the world outside.* And we have all the reason to think that the sequel of this processing, consciousness, has the same role.

This gives us fresh food for thought about a component of that world representation: the arrow of time. This arrow, the way mathematicians and physicists deal with it, is something impalpable, almost unearthly, something that belongs more on a calendar than on a map (see chapter

1). But in organisms with flesh and blood, and neurons, it rustles into life and becomes a feeling of time continuously flying from being past to being future. This is our most common conscious experience. However, where the past ends and the future begins is rather blurry; our "now" is born of a lifelong habit of fudging.

A major player in that conscious experience is our short-term memory—and I mean memory of the shortest sort, a remembering that lasts but seconds or fractions of a second. When we fleetingly recall the sequence of an array of letters flashed on a movie or computer screen, we use this kind of memory (psychologists call it *iconic memory*). The evidence that this memory is involved in our conscious experience of the passing of time comes from the neurological clinic, from studies on patients who had suffered brain lesions that left such memory losses in their wake. Consciousness then can be severely altered, even to the point of unawareness of the passing of time. It is difficult to imagine a loss of consciousness of that sort—from cradle to grave we are conscious of time being on the flow and take for granted that it will always be so.

Fortunately such extreme losses of memory are rare. But when they occur, they are a rich mine for the student of consciousness. The renowned neurologist Oliver Sacks relates a striking case, that of the musicologist Clive Wearing, who suffered a herpes virus brain infection that left him with a memory span of only seconds—the most devastating case of memory loss on record. Here is the story in lay terms, as told by the patient's wife. Her gripping words speak volumes:

> He did not seem to be able to retain any impression of anything for more than a blink. Indeed, if he did blink, his eyelids parted to reveal a new scene. The world before the blink was utterly forgotten. Each blink, each glance away and back, brought him an entirely new view. . . . It was as if every waking moment was the first waking moment. Clive was under the constant impression that he had just emerged from unconsciousness because he had no evidence in his own mind of ever being awake before. . . . "I haven't heard anything, seen anything, touched anything, smelled anything," he would say.

The psychologist George Miller, in a classic paper (1956), pointed out the existence of a memory system distinct from our long-term memory. This paper opened the door to what is now known as cognitive science. Miller's memory system has a limited capacity for holding information, in contrast to the long-term memory, whose capacity is seemingly limitless. By virtue of the short-term system, we can remember on the average about seven items—say, a seven-digit telephone number—without too much difficulty. But not many items more. It is this system that had been destroyed in Clive Wearing, and there are a few other reports in the neurological literature like that.

Where and how the short-term memory enters the machinery of consciousness is not known. But for the present discussion, let's take a reductionist's license. Assume that our sensory conscious experiences are made of brief unit conscious states, snapshots of the world outside, which are glued together by engrams of the short-term memory system. Then, as the engrams seamlessly link the unit conscious states together, we get a semblance of a streaming of consciousness with time.

So time as we feel it is something more than the disembodied mathematical entity it is in physics. Proust had it right. "An hour is not merely an hour," he wrote in *Remembrances of Things Past*, "it is a vase full of scents and sounds and projects and climates."

Unconscious Thinking

Our conscious experiences represent but a smidgen of the information entering our brain sensorium. Of the vast amounts of information that come in and get processed, only a minuscule part winds up as conscious states. And of the information processing itself, we have no conscious experience—the myriads of processings that at any moment go on in the brain are not accessible to our consciousness.

Nor, surprisingly, is much of our thinking. What little of that rises to the conscious surface are sensory images, mainly visual and auditory ones, and the latter are mostly related to language (phonological images). We call language the sequences of discrete sounds emitted by our vocal chords. But those really become a language only by virtue of the phonological images—it is the images that bear meaning, not the movements of our vocal chords. The phonological images by and large are not new information; they are mostly remembered sensory information that has been primped up a bit.

So if those on Mount Olympus ever trouble themselves to look at our glorified thinking, they may dismiss it with a laugh as mostly regurgitation, a behindhand sort of sensory perception.*

You may be inclined to think that rigorous logical reasoning is an exception, that surely mathematical thinking could not possibly be unconscious. Well, you would be wrong. Mathematicians themselves insist that some of their most profound thinking is unconscious. There is firsthand testimony in a fascinating book by the mathematician Jacques Hadamard, in which he offers his insights and those of two other distinguished mathematicians, George Polya and Norbert Wiener. All three stress that many of their cogitations are unconscious.

There is no shortage of other testimony of this sort, and quite a few go further back. Poincaré, for example, describes in one of his lectures how the crucial idea for one of his famous theorems (on the fuchsian function) suddenly came to him as he put his foot on the steps of a bus coming from the town of Coutances. And Gauss tells about a theorem he had unsuccessfully tried to prove for years, "Finally, two days ago, I succeeded, not on account of my painful efforts. . . . Like a sudden flash of lightning, the riddle happened to be solved. I myself cannot say what was the conducting thread which connected what I previously knew with what made my success possible"

So, yes, thinking can be largely unconscious, and I mean both, the thinking process and content. Not much, if anything, need ever bob up to the conscious surface. But whatever does, aside from visual images,

---

*Sigmund Freud already had a notion of this. "Only something which has once been a perception can become conscious," wrote the master of psychoanalysis in 1923. "[A]nything arising from within (apart from feelings) that seeks to become conscious must try to transform itself into external perception"—and this at a time when not much more was known of brain physiology than Broca's language center and the neuronal doctrine that Ramón y Cajal had announced just a few years before.

often takes on phonological form—and that's only befitting such a talkative ape.

From Piecemeal Information to Unity of Mind

We are certainly unconscious of the sensory-information processing itself, the myriads of processing and computational operations that go on at every instant in the wakeful brain. This at once raises the question of how the results of those operations get unified, as they somehow must when they give rise to a conscious experience. That experience, even the most fleeting one, is holistic. We see the meadow with its greens and browns and the weaving blades of grass, as a whole, and as if to give living proof of the oneness here, the hare sits snug in leaves and grass and laughs to see us pass.

Consider briefly again the situations on either side of the conscious surface. Below, things proceed analytically, as it were: the information from the world outside is splintered up into elements, a fragmentation that takes place in tens or hundreds of brain compartments quasi-simultaneously, each one making its own analytical world assessment. Above, it's the other way round: the elements are put together—combined one way or other and matched up. That synthesis and the holism of the corresponding conscious states are not just peculiarities of our visual experiences, but are traits of all our conscious sensory experiences: we hear the thrum of basses, timpani, trombones, and bassoons of a symphonic orchestra as one—and if the conductor has his way, on a good day even the pizzicato of violins is part of the oneness.

But what is it that pulls all those scattered sensory-information pieces together? What draws the results of the information processings in the various brain compartments into a whole? This is what among students of consciousness is known as the "binding problem." We will assume that Evolution solved it by standard neuronal communication, presupposing that conduction of information in digital form along axons or dendrites between the compartments is fast enough for the binding. At the ends of such information lines, the synapses, the conduction is typically slow. That dab of slothfulness can't be helped—it comes with the neuronal territory. The only way Evolution might have sped up things is by electrical synapses—junctions between neurons containing cell-to-cell membrane channels. Evolution used those synapses for synchronization of electrical signals, but I doubt that she would use them where fine control is needed. Although gaining speed, the system as a whole would stand much to lose in flexibility and integrative ability. In shaping her control systems—and the neuronal web is the control system par excellence—she traded off efficiency for flexibility.

That strategy of hers is as old as the hills. Control-system engineers are rediscovering it just now. But returning to the binding question, let's temper it and put it in terms of anatomically known brain tracts. I say tracts because the available evidence indicates that the binding involves vast numbers of neurons; one may thus expect sizable axon or dendrite bundles.

What tracts might draw the information-processing brain compartments together? Putting the question this way keeps it safely below the conscious surface (and us from becoming overambitious). Alas, our metaphor of information rising to the conscious surface now is wearing a bit thin. Evidently the bottom-up model for the brain sensorium only goes so far; at some point the flow of sensory information must become horizontal. Where?

Brain anatomy suggests several possibilities. Recall that layer V of the cortex contains the pyramidal cells, neurons with long axons that relay digital signals to distant cortical targets and, in turn, receive multiple (excitatory) inputs (figure 14.3). In particular, the pyramidal cells in a sheet-like structure called claustrum receive inputs from almost all cortical regions. These neurons, Crick and Koch have proposed, form loops of positive feedback, which keep the neurons firing much longer than they would on their own.

Aside from cortico-cortical feedback, there may be feedback via information lines going deeper into the brain . . . and further back in evolution. One tends to gravitate toward the cortex because it is here that much of the information processing culminating in consciousness seems to take place. But we shouldn't get too fixated on that precious grey of ours. It is possible, and from the evolutionary point of view quite plausi-

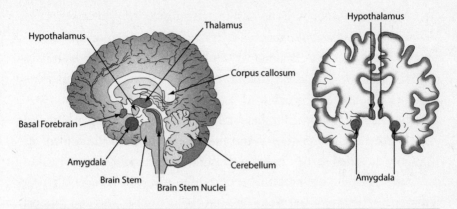

Figure 16.1. Human Brain. (*a*) in lateral view, (*b*) in cross section.

ble, that lower regions of the brain also take part in such processing. After all, birds have no cortex, and few would deny that they are aware of the world outside. Several areas in our brain are candidates here, especially the thalamus, which is a way station for nearly all incoming sensory information traffic (see figure 16.1). The thalamus neurons send that information on to our cortex, as we have seen, but they also receive information back from there, and the molecular biologist Gerald Edelman and his colleagues have put forth the hypothesis that this to and fro is part of a massive information feedback essential for consciousness.

Feelings and Emotions

The list of subcortical candidates does not end there. A number of other deep-set brain structures may be involved in consciousness; some lie in the hypothalamus, others in the amygdala, and even further down, in the brain stem (figure 16.1). Several of these structures were on the evolutionary scene long before the brain acquired its mantle of grey and are involved in conscious experiences of a more rudimentary sort: our joys and sorrows, our anger, fears, yearnings, and affections.

Such conscious experiences go under the umbrella of emotions, and I called them rudimentary because they lack the structured, sharp-image quality of our visual, auditory, and tactile conscious experiences. They have every appearance of being relics from the past—a remote evolutionary past when the neuronal web was still in its infancy. Recall that it was relatively late, a few hundred million years after the Grand Climacteric— about a fifth of the way down the post-climacteric traverse (figure 2.1)— that the neuron web had the necessary information capital to draw in earnest upon the information reserves in the structure of time (chapter 11). Only then had it acquired enough information-storing capacity and computer capability to compute the odds of future occurrences from past recurrences. Before, it was but a modest trellis ministering to a modest self—a self not yet bent on divining the future.

Things were a lot simpler then, and perhaps it is at the level of the relics from those times that our analysis of consciousness rightly ought to start. There is actually a fair amount of system-physiology knowledge of the relics. Indeed, it was here that physiologists originally cut their wisdom teeth and discovered the basic mechanisms that keep our body in balance. The quest was spearheaded by the nineteenth-century physiologist Claude Bérnard, who coined the term *homeostasis* for that balancing act. And what a stupendous act it is! Operating day and night and steered automatically by neuronal and chemical feedback, it manages to keep constant within narrow limits our body temperature, oxygen concentration in blood, and chemical profiles in tissue fluids, and it regulates our heart rate, the contractile state of blood vessels, breathing rate, and gut movements.

Our concern is mainly with the last four regulatory actions, because these often leave conscious experiences in their wake. Such experiences are of the relatively unstructured sort mentioned previously, but they can be quite intense and, if psychoanalysts have it right, leave long-lasting memory traces.

Take fear, for instance. Here the heart gets in the act—its beat may be slowed or even momentarily stopped in apprehension and fear. This is where the myth comes from that the heart is the seat of emotion. But in reality, the heart is just one of many integrants of the homeostatic complex—a mere component of that body machinery. Other components, though not with as much cachet, are the blood vessels and sweat glands in our skin; they are steered jointly with the heart by neurons in the brain stem. These neurons form coordinated sets sending digital signals to their targets when we experience apprehension or fear: one set sends the signals via the vagus nerves to the heart, commanding it to slow down or momentarily stop; another sends the signals via the cranial nerves to the skin of the face, making it blanch (or flush in embarrassment); while yet another sends them via the spinal tracts to the skin of the hands, making them clammy.

The signaling is largely automatic, but that doesn't mean that the brain is left in the dark about it. The brain receives information about the signals' homeostatic effects from sensory receptors located inside the body (*interoceptors*)—indeed, it gets round-the-clock information about the state of the whole homeostatic machinery. And just as the sensory receptors at the body periphery tell the brain about the outside world, the interoceptors do so about that inside world.

Our knowledge of how the brain represents that world still leaves much to be desired; studies on interoception, comparable to those on vision and touch done in the 1950s (chapter 15), were begun only in the past few years. But this much is known: interoception has its own dedicated brain centers, and that alone speaks volumes.

The most basic centers are largely concerned with the automatic steering of the homeostatic machinery. These are located in the brain stem. Others are in the cortex—in its cingulum and insula—and are no longer quite so automatic. At those centers information coming from the gut, heart, blood vessels, and erogenous zones gets processed— information about gut movement, throbbing of arteries, shudder, tickle, sexual excitation, and so on. The sensing here is mainly implemented by mechanoreceptors, and among them the ubiquitous Pacinian corpuscles once again get a piece of the action.

The part of this interoceptive information that rises to the conscious surface (a large part does not) produces a diffuse feeling of the state of our body. The nineteenth-century German physiologists had a word for it—*Gemeingefühl*, general feel—and regarded it as a sense. Alas, this has been forgotten, despite the importance of this sense staring us in the face. Indeed, we use it day in, day out to survey our body states and to gauge our well-being, and invoke it with every how-do-you-do.

It is this sense I had in mind when I spoke before of a relic from the past that might be a fitting target for a stab at the mechanisms of consciousness. Here, sensory information processing may not yet be as tangled and may not confuse the daylight out of us as in the much younger and more sophisticated sense of seeing. Only two other basic conscious experiences, the feelings of pleasure and pain, offer comparable prospects of simplicity.

The prospects already ebb with emotions. Take fear again. Here, memory and anticipation get into the act, entailing a lot more neuronal hardware. Indeed, recent magnetic resonance imaging studies indicate that, aside from the insula, the forebrain then becomes active; and this also holds for emotions like anxiety, despair, disgust, and anger. Recall that the forebrain has a hand in all sorts of anticipations, including those pertaining to our sense of seeing—it is here that cortical neurons compute the odds of recurrence of conscious experiences, based on past experiences (chapter 11).

Gut Feelings

Let me say a bit more about these ageless conscious experiences. They are not just felt internally; they tend to show in our deportment and mien. Such externalizations of feelings and emotions are important in their own right. Darwin was an early student of them, and in one of his lesser known opuses, "The Expression of Emotions in Man and Animals" (1872), he discourses at length on their evolutionary significance. Among these expressions, the facial ones are much in evidence in humans. Whether we are aware of it or not, we instinctively scan the faces of persons we meet for emotional signs—we follow their facial expressions with our eyes and read them for telltale signs of fear, anger, mistrust, and so on. That is the basis of our gut feelings, and for better or worse, they dominate our social intercourse.

Here is where the amygdala comes in again. This structure is a bit hidden away in the depth of the temporal lobe (figure 16.1), but it sits at the crossroads of cortical visual and auditory information traffic and plays a key role in the visual recognition of the facial expressions of fear and anger—and, by extension, in our social behavior. This was demonstrated by the neurologist Antonio Damasio. He and his colleagues carried out studies on patients suffering from Urbach-Wiethe disease, a genetic disorder in which the amygdalas on both sides were damaged and visibly replaced by calcium deposits. Such patients would no longer recognize the signs of fear and anger in people's faces, and they were behaviorally handicapped, perpetually vulnerable in their social interactions.

The scalpel of disease, when it cuts into tiny areas of the brain, can be quite revealing. Damasio has put it to good advantage in studying a number of other hidden brain functions. The interested reader is referred to his books (1999, 2003), which provide deep insights into the physiology of feelings and emotions, and are a delight to read.

Two Ways for Information to Get to the Promised Land

Let's go back now to our initial question, the problem of how to get from piecemeal sensory information to unity of mind. In the two major proposals on the table, loops of positive feedback, cortico-cortical or cortico-thalamical, would draw the results of the information processings in the various brain compartments into a whole. Such loops (and add now the cortico-subthalamic and deeper ones) would keep the pertinent neurons firing for tenths of a second—on the order of the duration of our conscious snapshots—considerably longer than they could on their own. Implicit in those proposals (and others of this sort) is the notion that such enhanced firing will have more impact on the information processing of the network—the loudest neuron assembly is heard over the din.

But there is another, more subtle, method to get impact: signal coherence. That method is more information economical than the brute-force one of positive information loops enhancing the running averages of signals—and is a natural when we have integrators like the pyramidal cells around, whose information throughput is inherently contingent on the timing of incoming signals.

Consider what happens at the input side of such a cell, its dendritic tree. Many axons converge on this strategic node of the network, each bringing its digital signals to bear. But it takes the concurrent (excitatory) inputs from several axons to satisfy the finicky cell—one input alone won't depolarize it enough to make it fire. To reach the firing threshold, the individual depolarizations produced by several axons must sum up. However, there is but a narrow window of time for such integration, because electrical charge decays rapidly in the membrane. It's a window of only one or two thousandths of a second (the membrane time constant). *As a result, signals arriving synchronously at the dendritic tree stand a better chance to get through to the next information-processing stage than signals arriving helter-skelter. Temporal coherence of signals, not the running average, here will carry the day.*

Thus, in the upper reaches of the cortex where the pyramidal cells call the shots, signal synchronicity is likely to be a significant factor in determining the throughput of the sensory-information stream. Indeed, it can be so decisive as to tip the scale on what kind of information gains access to consciousness.

The Virtues of Signal Synchrony as a Higher-Order Sensory Code

Let's backpedal a little to get a broader information perspective that includes coding, network coding. Bear in mind that here in the upper reaches of the sensory stream we are not dealing with run-of-the-mill communication networks, but with highly malleable, chameleonic networks—systems made of dynamically adjustable cooperative neuron assemblies (see chapter 15). The neurons in these assemblies can switch their cooperative allegiances in no time and be used over and over again in different coding combinations. It is in such networks that signal synchronicity shows its supple heft.

This brings us to the groundbreaking theoretical work of the physicist Christoph von der Malsburg. He has championed the idea of synchronicity encoding in neuron networks and showed analytically its advantages of cell economy and functional flexibility. And with that running start, it's but a hop to *cognitive* flexibility.

Synchronicity encoding, indeed, is an appealing idea, and if we continue this line of thought, it leads straight to the notion of coincidence detection by brain neurons. I can see quite a few neurophysiologists and computer scientists furrowing their brows here. Neurons are notoriously slow logic switches, and those in the cortex, and least of all the pyramidal cells, are no exception. So the question, and it is a crucial one, is, are nerve cells fast enough for that sort of detection? Could they do the necessary statistical computations in realistic times? By realistic I mean 20–30 thousandths of a second, or 100 at the outside, the times neurophysiological studies have shown to suffice for gestalt recognition.

The answer will depend on the type of correlation to be evaluated. Binary correlations would seem a no-go offhand; no nerve cell could even come near to bringing off the computations in time. But with higher-order correlations things decidedly perk up. Then, even a single event of n concurrent digital signals, with n large enough, say, 50 signals, gets to be statistically significant when set against random signals. The first arriving signals here will suffice for integration, and computation times may be in the realistic ballpark. Besides, if needed, time resolutions could be further improved by appropriate inhibitory circuitry, and von der Malsburg's analysis shows how this could be achieved at relatively low information cost.

In short, higher-order correlations may well be in the neurons' ken. So one ought to be on the lookout for congruous rhythms in signaling by neurons in different parts of the brain.

Signal Synchrony and Conscious Perception

Indeed, we needn't range far. The telltale signs of signal synchronicity are literally under our nose. Our olfactory system shows them prominently. Already in the 1940s, Lord Adrian, while recording from the olfactory system of rabbits, observed a regular electrical beat of 40,000 per second, the result of masses of neurons firing in synchrony. This didn't attract notice at that time—information theory was still in its infancy, not to speak of computers. But eventually signs of neuron synchrony would pop up all over the map. Following are some examples drawn from the literature of experimental neurophysiology of smell, hearing, and vision:

In the cat olfactory bulb and the olfactory cortex (front and rear), neurons fire in lockstep in response to familiar scents—and the beat is in the aforementioned frequency range. Auditory neurons in song birds respond to songs with characteristic digital-signal patterns, which are reproduced with high precision. So are the patterns of neurons in the auditory cortex of mammals in response to species-specific calls—in the owl monkey, for example, such patterns are reproduced with thousandths-of-a-second precision. Neurons in the retina and visual cortex of mammals fire synchronously in response to flickering light, faithfully following flicker frequencies of up to 50,000 per second, and the firings are in lockstep throughout several stages of information processing in the cortex.

All this goes to show that, sluggishness notwithstanding, neurons can operate with precise synchrony in the brain. They will do so along the ascending sensory stream all the way to the cortex and within the cortex, namely, across cortical columns, across entire cortical areas, and even across brain hemispheres.

This is all very well and good, but does that synchrony really serve the function we are so eager to foist upon it? Does it bring about the accouplement of the individual compartmentalized sensory-information processings in the cortex, or is it just an epiphenomenon? The doubting Thomas will demand evidence that the synchrony is necessary for perception, and he will not be put off by a lick and a promise. At the very least, he will want to be assured that the synchrony represents an internally coordinated phase-locking and not a mere stimulus-locking—a trivial synchrony, but by no means a remote possibility in a cortical network with bifurcating axons.

Concerns about stimulus-locking can be allayed. Synchronizations do occur unrelated to external stimulation. They occur even in the absence of such stimulation when memory is engaged. Our visions and dreams during sleep are cases in point. Episodes of synchronized neuron firing in our brain are then quite common, and the firing rates then depend solely on brain-intrinsic factors.

Neuronal Synchronization during Competition for Access to Consciousness

But there is no substitute for experiments in which the significant variables are controlled. The results of such experiments are now on hand and are most encouraging. They show a tight correlation between neuron synchronization and conscious perception. The neurophysiologist Wolf Singer and his colleagues demonstrated this in experiments on the visual systems of monkeys and cats. Monitoring the signals in the (unanesthetized) visual cortex, they showed that the neurons synchronize their electrical signals in physiologically meaningful conditions: when the animal's eyes were competing for access to consciousness.

In the case of our own eyes such competition, as psychologists have long known, can be a closely fought contest. It is a contest in which the winning eye takes it all. Our mind then tends to play tricks on us. Consider the drawing on the right. Chances are you see it as a cube. But fix your gaze on the edge deepest beneath the surface, and presto! the cube flips and that edge jumps out from the surface. It is not the object that flips, of course, but your mind.*

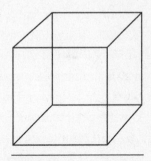

Figure 16.2.

Or take the gratings in figure 16.3. Look at a point midway between the two patterns of stripes and let yourself become a little cross-eyed. At some point the stripes become all vertical or all horizontal. Instead of seeing a plaid pattern, as you might expect if each eye had its way, all you see now is *one* pattern: either the vertical stripes or the horizontal

---

*This bistable illusion was described by the psychologist Louis Necker in 1832.

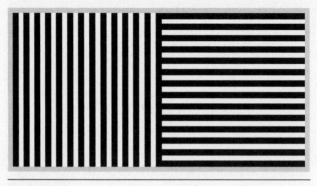

Figure 16.3.

ones. It is as if the left and right eye were vying for consciousness. And in a sense they are; their respective information-processing machineries are competing for such access, and it's either one or the other that wins the day.

Contests between neurons are routine in the brain web. All along the ascending information stream neurons at synaptic way stations try to assert their right of way. But this particular contest is special: it happens within earshot of the Promised Land, just a few hundredths of a second before the sensory information metamorphoses into consciousness. It is there that Singer and his colleagues chose to listen to the neuronal signals. They implanted multiple electrodes in the primary visual cortex of cats and monitored the firing of several neurons simultaneously, while patterns of stripes were shown to the animals' left and right eyes in conditions of competition. It was a competition that, by design, was permanently biased to one eye—surgery had made the animals, when young kittens, permanently cross-eyed. So each eye saw a pattern of stripes moving in opposite directions on a computer screen, bringing the crucial question down to the cortical neurons' firing when the stripe patterns were presented simultaneously to both eyes.

The answer was simple and can be summarized in a few words: the cortical neurons mediating the responses of the winning eye increased their synchronization, while those of the losing eye decreased theirs (see figure 16.4).

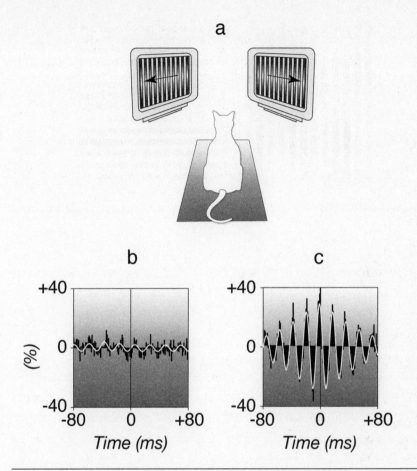

Figure 16.4. Neuron Synchronicity During Perceptual Competition. (*a*) Two mirrors were mounted in front of the cross-eyed cat, so that each eye saw a separate pattern of stripes moving in opposite directions on computer screens. Neuron firing was monitored in areas 17 and 18 of the primary visual cortex. The peaks of the correlograms (white lines) show the degree (percent) of synchrony of firing of the response of (*b*) the losing (left) and (c) the winning eye. (After Engel and Singer 2001.)

CHAPTER SEVENTEEN

Consciousness and
Quantum Information

So far in our journey we have seen the footprints of the Information Arrow fanning out all over the brain map. Some of them we were able to follow, despite bewildering neuronal tanglements, from the sensory outposts of the brain all the way to the cortex, to the doorsill of consciousness. Yet before we get too far with self-congratulations, it's good to remember that this was but the doorsill, and here the footprints came to an end, any material sign of them dissolving in unfathomable depths. Unfathomable, because we are completely at our wit's end to fit a physics process to the conscious afterclap of the Arrow. But if it's any comfort, it is also worth remembering that hereunto we have only looked at things through the lens of classical physics (contemporary neurophysiology is essentially classical physics). Thus our field of view has been rather narrow, and it is only natural to ask whether we'd fare better with the lens of quantum physics.

That is the question we'll ponder in this and the following chapters.

Boltzmann's World

Let me first make clear what I mean by classical physics. I mean linear physics, the world representation of interacting atoms and molecules, such as the practitioners of this science saw it in the nineteenth century and the beginning of the twentieth. Today's molecular biologists tacitly see the biological world in those terms. They may not say so, or not even be aware of it, but they are walking in the shoes of Ludwig Boltzmann, the great physicist of the nineteenth century. Indeed, the founders of molecular biology, like Francis Crick, Jim Watson, Jaques Monod, and Max Delbrück, would have felt quite at home with his notions of the world.

Boltzmann was actually an enthusiastic Darwinian. That is a sideline of his which is little known; I stumbled on it in the records of philosophical lectures of the University of Vienna. Boltzmann gave a series of lectures, drawing record audiences (even the Emperor Franz Josef was among his fans). Boltzmann discoursed on evolution, and he saw it beginning low down in the realm of molecules, as a mechanism whereby molecules will aggregate and turn into ever more complex structures, supramolecular conglomerates, stars, planets, and so on—and eventually into organisms capable of awareness and thought.

That was but a logical extension of his thoughts on the statistical behavior of molecules, for which he is better known. In his evolutionary conception things, although statistical at the level of the individual molecules, become more and more deterministic as the molecular structures grow in size. The sheer size of the structures, the sheer mass of their molecules, will ensure that the apple falls on Newton's head.

In Boltzmann's world, every event in nature was the result of accidental occurrences. The greater the number of individual occurrences from which a phenomenon is composed, the greater the probability that it had a deterministic character and the greater the likelihood that it followed definite laws. And when the probability that an event will occur is so high that from a human viewpoint it has become a certainty, we are entitled to speak of a law of nature.

But only then. The condition is that there be an immense number of molecules, an unimaginably large number of individual occurrences. Systems like that, despite their complexity and constituent-particle pandemonium, are entirely reliable and predictable: to know the positions and momenta of all the atoms and molecules at any given point in time was to know their positions and momenta at any other time.

And that about sums it up. Every natural event in that classical-physics view is quantitatively determined by the totality of circumstances or physical conditions at its occurrence. Every effect has its cause, and the whole is always equal to the sum of its parts. Nothing added, nothing subtracted (except information)—just the endless bouncing of molecules down the path of time.

Where the Quantum World May Come into Evolution's Ultimate Information Enterprise

Modern molecular biologists and neurophysiologists by and large have kept that deterministic worldview alive. It may seem odd, though, that they should be the standard-bearers of a philosophy that physicists abandoned decades ago. It's an old joke among physicists that they invented the deterministic-reductionist philosophy and taught it to the biologists, only to walk away from it themselves. But there really seemed to be no compelling reason for biologists to give up the cozy classical world; many processes in molecular biology and physiology are based on the actions of an enormous number of molecules, which follow the sober rules of classical physics closely enough.

However, that attitude will only go so far. Clearly, at sense organs like our eyes, the eerie quantum world cannot be ignored. The receptors in there, we have seen, are superb quantal machines (chapter 5). And when it comes to conscious perception or free will, it is hard to escape the suspicion that the pertinent neuronal processes are influenced, if not utterly dominated, by quantum mechanics. Granted, the transmission of information along the way to the brain, the digital signals traveling from the sensory periphery to the cortex and within the cortex, is reasonably well described in classical-physics terms. Not much of significance seems to be lost here if we sweep any quantum effects under the carpet. That part of neurophysiology is well accounted for by an equation the biophysicists Alan Hodgkin and Andrew Huxley formulated in the early 1950s—an equation based entirely on classical physics. Vast numbers of inorganic ions are at play here, and they follow Boltzmann mechanics. Even if we look at things at the level of the individual membrane channels—say, the filing of dehydrated potassium ions through the narrows of such channels (chapter 3)—they may be accounted for in terms of classical physics and chemistry.

However, higher up in the cortex, as we get to the information processing that ultimately gives rise to states of consciousness, things may no longer be so obliging. Neurons with large dendritic trees then enter the act, which integrate incoming electrical signals from hundreds or thousands of sensory lines (chapter 14), and crucial cognitive and behavioral decisions, whole motor patterns, may depend on just a handful of such neurons. Those decisions in turn depend on the vagaries of a few synaptic vesicles fusing with cell membranes. Add to that the modulations of these synaptic events by neuro-hormones, and we have a host of possibilities in which physiologically significant events might be ruled by the laws of quantum mechanics. Just as in the case of the interactions between photons and the electrons of rhodopsin inside the solid-state matrix of sensory cell membrane (chapter 5), so here crucial atomic interactions inside the matrix of neuronal cell membrane may be restricted to discrete quantum levels.

So there it is. We cannot leave out the bewildering quantum world from our naturalistic accounts of higher brain processes, lest we risk missing the mark. And as we grope our way through the web of the brain for clues to the mechanisms of consciousness, we must keep a weather eye out for quantum phenomena, even if they are strange, as quantum phenomena inevitably are.

"'Tis strange—but true; for truth is always strange, stranger than fiction," Byron said.

Quantum Information Waves

Our situation here—or call it quandary, if you wish—is really not so different from that in other searches in unexplored territories of science. It pays to dig for the deepest roots, and none are deeper than those of quantum mechanics.

To illustrate this point, I'll take a page from the history of chemistry, of how we came to understand what holds molecules together. When in the 1920s the physicists Walter Heitler and Fritz London attacked that problem, the forces between molecules were still unknown territory. A hazy notion based on chemical reactions was all there was—chemistry was limping along on metaphors.

But that situation changed almost overnight as Heitler and London brought quantum mechanics to bear on the problem. In fact, that was the first time that quantum mechanics, namely Schrödinger's wave function (see box 5.1), was applied to a problem outside physics. The yield for chemistry was huge: a rational framework for chemical bonds and reactions. It is hard to imagine how else modern chemistry could have gotten off the ground, or how molecular biology could have—genetic-information transfer would have been inconceivable.

To be sure, bringing quantum mechanics into brain physiology will be more difficult. Unlike Heitler and London, we have no general theory to guide us here—just Evolution's awesome lottery.

However, we are not entirely in the dark. Quantum bits, not macroscopic bits, are nature's universal information currency. On nature's ground floor—and that includes the atomic level of molecules—a wide variety of interactions provides a virtually inexhaustible source of bits of this sort, ready-made for high-speed information processing.

Evolution has not overlooked those opportunities in her great information enterprise. At the brain's sensory periphery, we have seen, she exploited them to the hilt, including coherency of quantum waves, So if we keep a sharp lookout elsewhere in that immense web for high-speed information processing, she may throw us a hint on where to do our digging. Those ghostly waves are really not as unruly as they may seem. Schrödinger tamed them quite a bit, and they obligingly follow his mathematics as probability waves. If we put those waves under the information loupe, they shed their mantle of mystery, or at least one of the mantles. Then they show themselves for what they are: information waves. They tell us where in space the particle is likely to turn up. The amplitude of the wave imparts that information: in places where the amplitude is high, the likelihood is high; and where the amplitude is low, the likelihood is low (see figure 17.1).

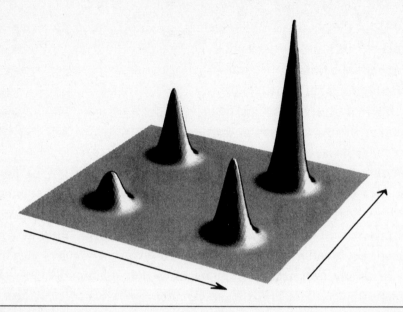

Figure 17.1. Quantum Waves. Representation of a particle's wave-function's many possibilities (density matrix). The wave amplitude is largest at the locations the particle is most likely to be found. (After Joos 2010.)

The Vanishing Act

That is all very well and good. At the elementary level, matter proves to have a wavelike character, and the waves are mathematically tractable. So long as one stays at that level, the mathematical laws governing the quantum world are remarkably precise. But when one gets to the macroscopic everyday level, there is a sudden change of key, and this happens as soon as one gets to the microscopic/macroscopic borderland. Then, all of a sudden, the quantum wave vanishes.

No, you didn't misread. Careful experiments bear witness to the disappearance of the superposed quantum states. And this is no sleight of hand, nor sleight of mind. Precise measurements with a variety of quantum particles prove the reality of the occurrence, though it is an act that would have made a Houdini green with envy. So long as we leave a quantum system alone, it behaves like a wave, all its informational possibilities following Schrödinger's equation in a mathematically rigorous manner.

But as soon as we perform a measurement on it, or only so much as observe it, the elves hearken to a different piper: the previously large number of possibilities coexisting in superposition are abruptly reduced to one (see figure 17.2).

It is as if the act of observation or measurement caused the many potential outcomes to crystallize into a single macroscopic reality. Quantum physicists call this the *collapse of the wave function*.

There still is no entirely satisfactory explanation for the collapse, but the notion has proven very useful in all sorts of mathematical dealings with the microscopic/macroscopic borderland. The founders of quantum theory—Schrödinger, Heisenberg, Bohr, Dirac, von Neumann, Eugene Wigner—all have availed themselves of the notion in one way or other. But as to the question of what mechanism causes the wave to collapse, even today, after seven decades, one can but speak ex cathedra ignorantiae.

However, there are several interesting proposals on the table, which I will now discuss insofar as they bear on consciousness.

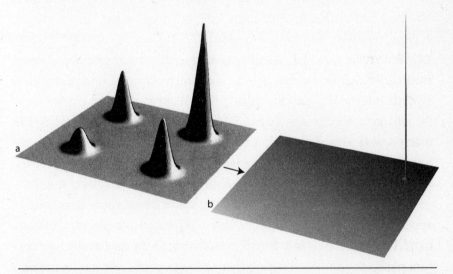

Figure 17.2. Collapse of the Quantum Wave. The many superposed coexisting possibilities before observation (*a*) are reduced to one when a particle is actually observed (*b*). At that location, the probability of finding a particle then surges to 100 percent, while it drops to zero at all other locations.

Are Quantum Waves and Consciousness Entangled?

I'll start with an old proposal by Wigner, which historically set the tone. This proposal by and large reflects the sentiment of several other founders of quantum mechanics. But Wigner takes the bull by the horns and links the wave collapse outright to consciousness. To him the collapsing of the wave function during observation suggests that the wave gets entangled with a conscious being, and that this entanglement somehow causes it to collapse.

That line of thought can be found in a number of later proposals. The most concrete, and most daring, is by the mathematician and physicist Roger Penrose. It is part of a wider effort of his to bridge the gap between the quantum and macroscopic worlds, a chasm that has stubbornly defied two generations of physicists. To fully do the proposal justice, we need to examine it in that context.

Penrose endeavors to bridge the gap by merging general relativity—and there is no one who has contributed more to our understanding of general relativity, except Einstein himself—with quantum mechanics. His core idea is that gravity causes the quantum wave to collapse. In general-relativity theory, the spacetime curvature is ruled by gravity, and Penrose argues that when spacetime in the brain gets sufficiently warped by gravity, namely, a warping on the scale of one graviton, a nonlinear instability sets in, causing the collapse.

This puts the entanglement squarely in the brain, and Penrose ventures even further, assigning it to a particular cell material: tubulin. He and the anesthesiologist Stuart Hameroff hypothesize that this protein, which forms tubes—the "microtubules"—inside cells, causes the wave collapse through a cooperative interaction of its subunits in quantum superposition, as the number of such subunits accrues to a critical threshold.* This is the aforementioned gravity-related threshold which, trans-

---

*The tubulin subunits are arrayed in columns, forming tubes of about 25 millionths of a millimeter in diameter, the microtubules. These are common fixtures inside cells, but what arrests one's notice is an electron strategically located inside a nonpolar pocket of the tubulin subunit, which can shift from one position to another, causing

lated to molecular terms, would take a mass of about a billion tubulin molecules in superposition to reach it in half a second. We'll return to this hypothesis further on and weigh its feasibility, after taking a long, hard look at the constraints that nature imposes.

Irretrievable Quantum Information Losses: Decoherence

Another proposal, spearheaded by the physicist Dieter Zeh, bypasses conscious beings altogether. Zeh offers an entirely different perspective on wave-function collapse, where the environment itself sees to the collapse and consciousness has no special role in it. Here the cumulative action of the particles swarming in our environment—the photons from the sun, the photons from the deeper regions of the cosmos, the molecules in the air, and so on—would cause the collapse. Such particles won't noticeably dent a large object, and we have seen why (chapter 5). But that doesn't mean they leave its wave function unscathed. In fact, they continually nudge it, and by their sheer numbers ensure that any wave coherence, any superposed quantum state in the object, gets snuffed out.

What gets dented here are the energy fields of the object's constituent particles. They are warped by the environmental particles impinging on them. This continual warping causes the orderly sequences in an object's quantum waves, of crest followed by trough followed by crest, to be jumbled up—or in the physicist's lingo, the waves *decohere* (see figure 17.3).

Now, it pays to look at this process from the information angle—it always does when we want to get to the bottom of things. Recall that wave functions are information waves; Ψ itself is information, potential information. If we use our information loupe, the decoherence process bares its essence: the wave function loses information—the information

---

the subunit to switch between two molecular conformations. This is the feature Hameroff and Penrose seize upon. They hypothesize that these conformational states occur in quantum superposition until the moment of collapse, an "orchestrated" self-collapse through a cooperative accrual of coherent subunits.

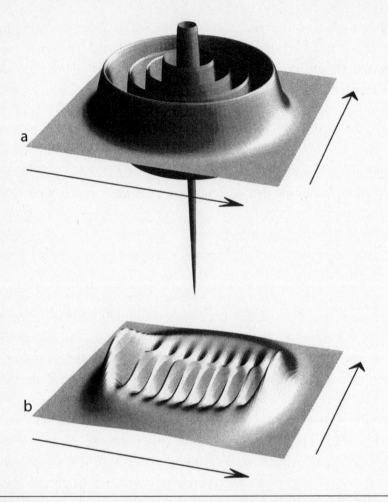

Figure 17.3. Decoherence. The wave function (Wigner function) of a quantum-coherent system, a harmonic oscillator, before (*a*) and after decoherence (*b*). (From Joos 2010.)

in an object's wave functions is continually being nibbled at and, bit by quantum bit, is carried into the environment, never to be recovered.

We are dealing here with an information loss at the most elementary level. You will probably be more familiar with another irrevocable information loss, which takes place much higher up, at the molecular level. That kind of loss is through heat transfer and is the basis of classical thermodynamics. But that kind we can prevent (or at least slow down significantly), whereas the losses by decoherence we cannot; there is no way of completely shielding an object from environmental quantum particles. Even if we could block out the photons, we couldn't block the neutrinos—even in the darkness of intergalactic space the object's wave functions would be nudged by the microwave photons left over from the Big Bang (see chapter 2).

In this sense, decoherence is more fundamental than the classical-thermodynamics process. And the fact that what gets lost are quantum bits of information, rather than classical Shannon bits, underscores this point. Indeed, decoherence may well be the most fundamental irreversible process in nature.

Why Our Weltanschauung Is So Narrow: An Unexpected Lesson for Philosophy

The decoherence proposal has come a long way in the past few years and, in the hands of Zeh, Erich Joos, Wojciech Zurek, Murray Gell-Mann, Jim Hartle, and others, has grown into a comprehensive theory. It is a theory that leaves hardly anything untouched, and not just physics. Its reach goes well beyond, to what once was the domain of philosophers: epistemology, the theory of knowledge.

This brings us back to our center of interest, the issue of whether the quantum wave gets entangled with consciousness. In that regard, the decoherence theory gives us a freedom of thought the founders of quantum theory never had. To many of them, such an entanglement seemed a Hobson's choice (e.g., Wigner's proposal). What else was one to make of the fact that the wave function evaporated whenever a human observer got in on the act? Somewhere between the microscopic and macroscopic the abracadabra had to happen, and the conscious brain seemed just the place for it. Well, as Aristotle famously said, "There's many a slip 'twixt cup and lip."

Decoherence theory changed all that. What once seemed a dire necessity is no longer so. Quantum decoherence goes on at immense speeds in our earthly atmosphere, and wave function collapse ordinarily is over long before a human observer or measuring device intrudes.

Or look at it the information way. The environment then suddenly metamorphoses from an innocuous particle hinterland into a voracious information sinkhole. And as it gobbles up those immense amounts of bits coming onto us from all points of the compass, it drastically curtails our view of the world outside—much more so than all our sensory systems combined. It is nature herself who acts as censor here and sees to it that we have but the narrowest perception of reality.

No human observer is necessary for the wave-function collapse. In a sense it is the environment that performs the observations and measurements here—and it does so in almost no time at all—as it turns all the coexisting quantum states into one single macroscopic counterpart (see figure 17.2b).

That is an epistemological twist philosophers hardly expected—and to tell the truth, neither did physicists. But it's the former who are the professed mavens of epistemology and, historically, its custodians. Well, no longer. Epistemology has come of age and joined the ranks of science.

On the Possibility of Quantum Coherence in the Brain

But philosophy aside, to the student of consciousness, decoherence theory provides a welcome breath of fresh air. It cuts the ground from under the notion of an obligatory entanglement of the quantum world with our mind and at one stroke makes hypotheses about consciousness based on wave-function collapse less appealing, if not unnecessary.

In any event, to those intent on quantum-mechanical hypotheses of consciousness, it offers sobering constraints, holding exuberance in rein. I was not exaggerating when I said that decoherence ordinarily takes almost no time. In our earthly atmosphere the timescales of decoherence ordinarily are unimaginably short—on the order of 10^{-13} seconds or less. And if they are that short where air molecules are the main decoherent agents, they will be still shorter in an environment like the brain, swarming with water molecules and inorganic ions; the calculated decoherence times here range from 10^{-13} to 10^{-20} seconds—a far cry from the physiological timescales of consciousness, 10^{-2} to 10^{-1} seconds.

This imposes harsh demands on the Penrose proposal, whose central tenet is the existence of a quantum-coherent brain material, a material that, before consciousness sets in, will maintain quantum states in superposition. The question thus is—and this is the crux—will the postulated material hold up under the onslaught of local water molecules and ions and delay the natural course of things by some 11 orders of magnitude?

It is not just the water and ions in the cell interiors one must reckon with, but also the water and ions outside the cells. Their presence in the intercellular space imposes additional demands, spatial ones. Our conscious experiences, we have seen, are multicellular affairs; they involve vast numbers of brain cells—the perception of even a minimalist gestalt entails the participation of sizable groups of cortical cells (chapter 15). The postulated material, therefore, must extend beyond the cell boundaries and bridge the intercellular spaces, lest its quantum coherence be nipped in the bud.

That doesn't bode well for microtubules, the candidates Penrose and Hameroff have advanced. These structures are ubiquitous inside cells (all sorts of cells, not just neurons)—they crisscross the cell interior but don't traverse the cell membrane. Thus, they don't provide the necessary intercellular continuity, a shortfall looming all the larger in an orchestrated recruitment of coherent tubulin subunits such as Penrose and Hameroff propose. And it is no good to invoke gap junctions here, and better not try. Their intercellular channels offer no shelter against water and inorganic ions; they are full of them themselves.

CHAPTER EIGHTEEN

Molecular Quantum Information Processing and Quantum Computing

Having ended the previous chapter on a downbeat note, I wish to emphasize that my concern is with tubulin, not with the *general* idea of a quantum-coherent brain material. Such a material would have much to offer in its own right, regardless of consciousness. The brain is not so much about consciousness as it is, and cannot escape being, about computing—and above all, parallel computing. To perform myriads of parallel computations is par for the course for its neuron web (chapter 14), and for this reason alone a quantum-coherent neuronal material would be a tremendous boon. I wouldn't be in the least surprised if Evolution, ever on the lookout for information economy, had come up with materials that stay quantum-coherent long enough for physiologically meaningful computations. Quantum computation is a natural for parallel computation and is the perfect information bargain.

Quantum computation may not sound like much—it may give the impression of being just another method of computation. Well, that's what it is, but it is also a good deal more. It is a method that uses the strange rules of the quantum world but provides undreamed-of power. A machine playing by these rules will look nothing like the one that sits on your desk. It will be rigged with novel but surprisingly down-to-earth hardware like that sketched out in the following pages. But the chances are, it will be rigged more like your brain.

Quantum Computers

Quantum computers have an aura of mystery, though that's hardly warranted. Such an aura should be reserved for things we analytically cannot fathom. Granted, a definition of quantum computers is not easy because those machines can take on many physical forms. But if we cut across the morphological trivia, we can rustle up a definition in a few words: *a quantum computer is a device that extracts information from quantum waves in superposition.*

So those haunting poltergeists in Ψ have their use after all, But I doubt that Schrödinger would have thought it would ever be *that* way. But so it goes. Eventually, once quantum theory had gotten over its long birth pangs, in the 1980s Richard Feynman, Paul Benioff, and David Deutsch advanced the notion that those superposed quantum states might be harnessed to store bits of information. Just as in a standard digital computer a set of capacitors obeying the rules of classical physics can store information, so in principle could a set of atoms—say, a string of hydrogen atoms—obeying the rules of quantum mechanics. In the digital computer the voltage between the plates of the capacitors represents a bit of information. A capacitor here encodes one of two alternative states: charged or uncharged—which in computer language corresponds to a 1 and a 0. In the quantum counterpart, an atom in the excited electronic state encodes a 1 and in its electronic ground state, a 0.

So far the machine looks rather tame. It seems but an extension of Turing's Universal Machine (chapter 12), and in a purely mathematical sense, it is. But not in the physics sense, and not by a long shot. As soon as we consider its inner makeup, we realize what is driving it: atoms, electrons, nuclear spins, or just plain photons, in quivering superposition. Ephemeral as they are, those are the machine's nuts and bolts—its unit-computing elements. And thanks to them, the machine can be in many states at once! That is, in as many states as it has elements in superposition.

The Advantages of Quantum Computing

You may ask, how many of such states are we dealing with here? The answer depends on the machine, and we will see shortly why we must hedge our bets. But since the discrete character of quantum mechanics matches that of digital information processing, we may, for the sake of comparison, mark off the information limits: a quantum computer with n quantum bits can be in up to 2^n different states at once, whereas a standard digital computer can only be in *one* of 2^n states at any one time.

In short, *a quantum computer can in principle be in myriads of states simultaneously and hence do myriads of operations simultaneously. It is a natural parallel computer.*

So, there is no aura of mystery here—just raw power; indeed, unmatched power. But before we take its measure, let me clarify one thing. When I said the computer can do myriads of operations simultaneously, I meant one single processing unit. And to leave no doubts I spell it out completely: one processing unit here does with one fell swoop what would require myriads of processing units of a standard computer. Such a unit would be able to solve problems one couldn't even dream of tackling with today's digital computers. Take, for example, factorization, a bread-and-butter operation of those guarding (or breaking) our banking codes. Factorization is essentially the reverse of multiplication, but harder: one is given a number (the input number) and asked to find two smaller numbers (factors) that multiplied together yield the input number. The obvious way is to divide that number by all possible factors, starting with 2 and continuing with every odd number, until one of them divides the input exactly. At least one of the factors here can be no larger than the input's square root (provided the input is not a prime), and this affords one an estimate of the computing time. For an input number of, say, 10 digits, that time isn't inordinately long; this sort of problem is readily manageable by standard computers. However, the time increases exponentially with the number of digits, and as we get to the 100-digit range, it becomes hopeless (the present security codes for our bank and credit cards use on the order of 250 digits). One would need hundreds of state-of-the-art digital computers linked up by Internet to perform the trillions of operations necessary for the factoring. On the other hand, a single quantum computer could do it all in the blink of an eye.*

Or take a problem of more immediate biological interest, the generation of virtual computer images. That is what a truly universal computer excels at. A standard digital computer may swing a reasonable semblance; the information for its operation resides in the states of the computer "head," and it is here that the structure of the object to be imaged is mapped. With today's digital computers it is possible to render images of structures having 10^{12} sites. But this requires a whole network of such computers. With better computer hardware, one might perhaps push the resolution to 10^{16} sites. But that is about the most sanguine expectation we might seriously entertain. On the other hand, for a quantum computer, all this would be child's play; with so many more states to map on, one might resolve even quantum particles in action.

Quantum Computers in the Real: Computing with Atoms

This much for quantum computation in the abstract. And not so long ago we couldn't have said much more; theory was well ahead of experiment. Not that this is unusual in science, but rarely has the disparity been so extreme and theory and practice so hopelessly out of sync. And for good reasons. We are up against nature's most unforgiving process here, decoherence, which in our environment almost instantaneously spoils any quantum superposition (chapter 17). Pity the engineers who are saddled with the task of building a quantum computer. They must insulate the inner workings of the computer from the environment, or else its superposed states will immediately end up on the

---

*This bodes ill for our bank and credit cards. Even a number of 10^{200} digits could be factored in a matter of seconds with the aid of a quantum-computer algorithm discovered in 1994 by Peter Shor.

trash heap; yet they must leave the inner workings accessible for information loading and read-out. It's damned-if-you-do-and-damned-if-you-don't. No wonder there are so few quantum mechanics.

But things are changing and perking up. This began in 1994 when the physicists David Wineland and Christopher Monroe built an ingenious device to stave off decoherence. It did so only for a twinkling, but even so it was a triumph, considering that in ordinary environment it takes barely 10^{-13} seconds for coherent quantum waves to collapse. Wineland's and Monroe's device was essentially an ion trap: a strong electromagnetic field constraining a beryllium ion in three dimensions. In that hammerlock grip the ion could be efficiently zapped with photons supplied by a laser beam. The ion had a net positive charge—the original atom had been stripped of one electron—which is why it could be trapped. But it still had one electron remaining in its outermost shell, and this was the laser target. With every hit the target was knocked from one energy-quantum state to another or into a mode where both states were in superposition. These various states thus encoded bits of quantum information, allowing the system to perform operations of quantum logic.*

At about the same time, the quantum information theorist Seth Lloyd and the experimental physicists Jeff Kimble and his colleagues set out on a different path. They trapped photons between two highly polished mirrors and forced them to bounce back and forth, while cesium atoms were dripped into the space between the mirrors at right angles. The mirrors were only millimeters apart, so a photon was likely to be captured and released many times over by a cesium atom. As in the ion trap above, only two electron-energy states are involved in that interaction (and in the encoding of qubits). The photons would eventually escape the trap and their quantum states be read out by polarization measurements. In this case the photons rather than the atoms were the ones encoding the bits of quantum information. They were bits on the fly, as it were—but good enough for quantum-logic operations.

---

*The photons here caused the finer energy modes to change, modes depending on the alignment between the spin of the electron and the nucleus of the beryllium ion. These internal energy states encoded one qubit, and two vibrational modes, the other qubit.

And quantum logic is the name of the game. True, those were minimal operations—the computers were only rudimentary prototypes. But it was a beginning, and efforts are currently under way to scale things up. Large ion-trap arrays are in the works in Wineland's lab, which at the time of this writing had raised the processing capacity to 10 ion-qubits; and in Kimble's lab, models combining ion-trap and photon-trap technologies are on the drawing board. It remains to be seen how far things can be pushed to scale up operations.

Quantum Computing with Atomic Nuclei

A different attack on the quantum-computer problem was launched in 1997 by the physicists Neil Gershenfeld and Isaac Chuang, and by David Cory and his colleagues. They set out to harness the quantum information inherent in molecules in liquids. At first sight that may seem a quixotic pursuit—we have seen what a shellacking quantum waves receive in water! But the true and steadfast scientist occasionally is granted a reprieve.

Indeed, it turned out that a molecule with an innate aptitude for quantum computation is no rara avis. Nor is it all that difficult to spot; the bird gives itself away by its peculiar behavior, or better said, the behavior of some of its atomic nuclei. Like other particles, these nuclei behave like spinning tops—they spin around an axis and wobble a bit. But they are highly sensitive to external electromagnetic fields, and that is what the two teams mentioned above took advantage of, and what ultimately allowed them, against all odds, to tap the nuclei's quantum information.

The telltale signs of this nuclear proclivity can be picked up by a standard technique of physical chemistry: nuclear-magnetic-resonance spectroscopy. This method had already proven its worth many times over, enriching science as well as technology (MRI, the magnetic resonance imaging used in hospitals for examining people's innards, is perhaps its most widely known offshoot). Chemists and biophysicists have employed the technique for decades to study the structure of molecules in solution, and much of what we know about such molecules, especially biomolecules, is owed to their imaginative use of the technique. But little did they know that they had been quantum computing all along.

How does one quantum compute with nuclei? How does one maneuver those wobbly spinning tops into relinquishing some of their closely held information? That takes some savvy—one doesn't get very far with brute force in information matters.

Let's put a nucleus, say, that of the hydrogen atom (^{1}H), under the information loupe. This particle has a suitable spin for nuclear-magnetic-resonance work. Its rate of rotation is the same as that of the electron—spin ½ in the parlance of the trade.* That may seem surprising because the hydrogen nucleus has a much larger mass than the electron, but the rate of spin is an intrinsic property of an elementary particle, much like its mass or charge is—the rotation rate of every electron in the universe is the same. And as a general piece of information, let me add that all elementary particles of matter (neutrinos, quarks, muons, tau) have spin ½, whereas the elementary particles of force (photons, gluons, weak gauge bosons) have spin 1.

The hydrogen nucleus has a positive charge (it is a proton) and, like any moving electrical particle, it will generate a magnetic moment. It thus tends to align itself along the direction of an externally applied magnetic field, adopting an orientation either parallel or antiparallel to it (see figure 18.1a). These two nuclear orientations correspond to two alternative quantum states (the parallel orientation has less energy than the antiparallel one). They constitute natural quantum bits.

There are more bits to be had if we add an oscillating electromagnetic field to that backdrop. Then, at the appropriate frequency (*resonance frequency*), the pulses will nudge the spinning nucleus ever so slightly during the same point of gyration, and if we control the angle of the pulses and the time they are applied, the nucleus can be tilted sideways or flipped over completely, adding more quantum states in superposition (see figure 18.1b).

So a thimbleful of molecules goes a long way in providing quantum bits. Even smallish molecules with just a few nuclei amenable to being

---

*Meaning that the angular momentum of the particle from its spin is h/2.

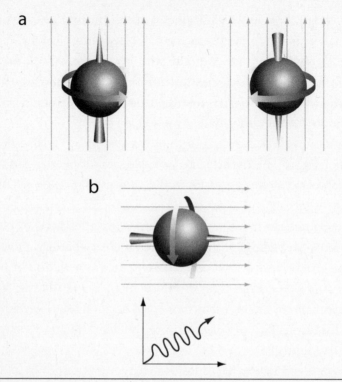

Figure 18.1. Manipulating a Hydrogen Nucleus. (*a*) The spinning nucleus aligns itself along the direction of an external magnetic field, ending up gyrating either counterclockwise (*spin-up*) or clockwise (*spin-down*). Spin-up conventionally registers the logical state O and spin-down, state 1. (*b*) An oscillating electromagnetic field applied at an angle tilts the nucleus sideways (*spin-sideways*) or flips it over. Spin-sideways is a quantum state that registers 0 and 1 at the same time.

reoriented may be useful. But quantum bits don't a computer make— no matter how many. The prerequisite here, and this holds for a quantum as well as a classical computer, is the capacity for implementing reversible unitary transformations on the information input (chapter 12). That is what betokens computerness—and what's sauce for the goose is sauce for the gander.

Natural Quantum-Computer Games

I have been harping on the similarities between classical and quantum computers here—perhaps a bit too much, as the similarities will only go so far. Classical computing is by and large intuitively clear. Quantum computing, on the other hand, has inherited all the nonintuitive aspects of quantum mechanics. To start with, the information units, the bits, encode different things. In a classical computer, a bit encodes either a false or a true—a 0 or a 1 in conventional notation—whereas in a quantum computer, it encodes a 1, a 0, or a superposition of these (figure 18.1). The unitary transformations in the two cases are necessarily cast in a different mold. I mean this literally, because the physical devices implementing the operations—the *logical gates* in the language of the trade—are structurally very different.

Consider the operations of a classical computer. These are intuitively clear; they are the distillation of sound reasoning, the implementation of the "Laws of Thought," which the logician George Boole set forth 150 years ago. The simplest operation is what is called a *NOT* operation in Boolean logic. It is merely a flipping of bits: true becomes false, and false becomes true. And with two more operations we can close our list: *COPY* and *AND*. *COPY* makes the value of a second bit the same as the first.

AND makes the value of a third bit equal to 1 if, and only if, the input bits are both 1; otherwise it makes it equal to 0. *COPY*, like *NOT*, is a linear operation—the information output reflects the value of a single input—whereas *AND* is nonlinear. These three operations form the computer basics; they are all that is needed to perform any task of logic, including complex mathematics. The gates implementing the operations go under the name of *NOT-*, *COPY-*, and *AND* gates. They are the backbone of a classical universal computer.

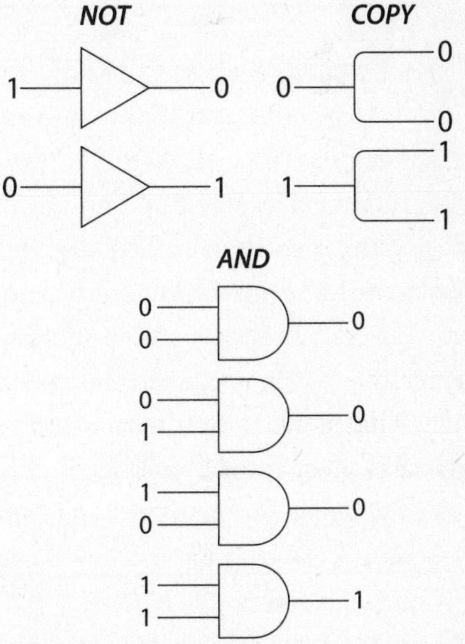

Figure 18.2. Logic Gates.

What are the equivalents in a quantum computer? Here the operations no longer have that intuitive feel. Nothing in our daily experience prepares us for bits being flipped halfway and the superposition of the corresponding states, which is sine qua non for a quantum computer. That sort of flipping is the purview of a special gate: the Hadamard gate, H gate for short—named after the mathematician we met in chapter 16. It brings about a unique transformation: it turns a spin-up or a spin-down 180 degrees around a slanting axis—the most elementary of unitary transformations (which Hadamard discovered). Two such transformations in sequence will go full circle, returning the superpositions produced by the flipping of bits to the original state. Hence, two H gates in series are the equivalent of a *NOT* gate (see box 18.1).

Box 18.1.

The Hadamard gate —[H]— A one-qubit gate that produces an equal superposition of two quantum states, the most fundamental operation that can be performed on a quantum bit. Two Hadamard gates in series —[H]—[H]— are the equivalent of —▷— , a classic NOT gate.

But what of the equivalent of the *AND* gate, the one turning the indispensable nonlinear trick? Conventional wisdom had it that, after the fashion of a classical computer, this would require a three-bit gate—a simultaneous interaction among three quantum states. That would be something fiendishly difficult to engineer, and to the would-be builders of quantum computers, it seemed an insurmountable hurdle.

Actually, it was the *pons asinorum*. All those worries turned out to be misplaced. The answer to the nonlinearity problem was simpler than anybody thought. Best of all, it has a relatively low information price. With the benefit of hindsight, it is easy to see why: nonlinearity can be had wholesale in the microscopic world—colliding atoms and a host of smaller interacting particles perform nonlinear operations as a matter of course, offering a rich quarry waiting to be mined for quantum logic. But that would only sink in later. The theoretical breakthrough came in 1995 when Deutsch, Adriano Baranco, and Artur Eckert, and independently Lloyd, showed that a two-qubit gate satisfied the requirements for reversible quantum logic. A three-qubit gate was superfluous—or worse, redundant.

The circle of students of quantum computing heaved a sigh of relief. A two-bit quantum gate was something doable. One could think of a number of naturally occurring molecules, especially organic ones, that might be pressed into service here, and with the right kind of electromagnetic prodding and a little bit of luck, the gate's molecular pieces might fall right into place.

How to Rig a Molecular Quantum Computer

This is where the nuclear-magnetic-resonance experiments came in. We saw how the nuclear spins inside molecules can be manipulated and how this yields superposed quantum states suitable for encoding quantum bits. Here we get down to the molecular nitty-gritty, the genesis of quantum-logic gates.

We'll use a simple model, dibromothiophene (see figure 18.3) This is one of the molecules Gerschenfeld and Chuang originally worked with. It is rather plain, as organic molecules go. But it has what it takes for quantum logic: two hydrogen nuclei, each with two mutually exclusive magnetic orientations that can encode a quantum bit; moreover,

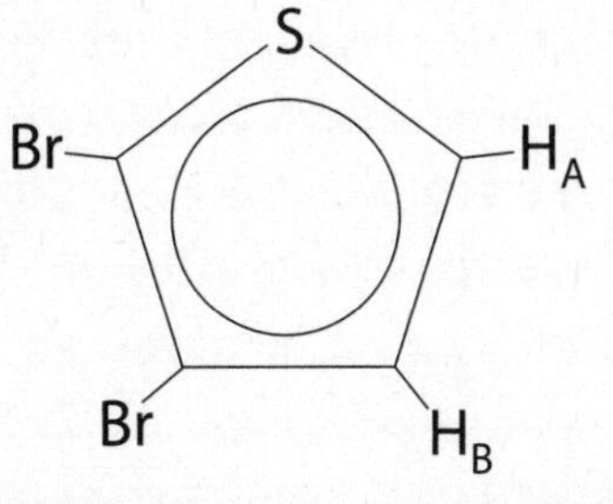

Figure 18.3. Dibromothiophene. The two hydrogen nuclei (H_A, H_B) have slightly different nuclear-magnetic resonance because of their different neighbors (S, Br), and so can be addressed separately. Their respective different orientations encode one qubit, and appropriate external electromagnetic pulses produce superposition states of the qubit, bringing about a Hadamard transformation.

the molecule can perform Hadamard transformations on the nuclei when prodded with electromagnetic pulses.

But neither the encoding nor the Hadamard transformations by themselves are good enough for computing. Two Hadamard transformations, however perfectly they may flip and slant those nuclear spins, won't meet the demands of reversible quantum logic. This is no easy thing. It calls for a nonlinear interplay between the two transformations, namely, an interaction that, after the fashion of a classical *AND* gate, makes one of the transformations conditional on the information state generated by the other.

Fortunately one needn't trouble one's head about how to hack such an interaction. The bromothiophene molecule sees to that by itself. This is one of those natural nonlinear operations in the atom world that I spoke of, and it is where the bromine and sulfur atoms, the immediate neighbors of the hydrogen nuclei here, come in. The *local* magnetic influences these atoms exert on the two nuclei are slightly different—a discrepancy translating into a difference in nuclear magnetic resonance frequency.* The two nuclei can therefore be addressed separately, opening the door to a semicontrolled, sequential flipping of spins.

And that is the ticket to the soup. The computer operator really doesn't have to do much more than mete out the two resonance frequencies with the right timing. That will set in motion the molecular wheelworks and give the run of the whole quantum cabal: two Hadamard transformations that are so intertied, the second transformation is conditional on the quantum state produced by the first. And all that results automatically. The two operations are in lockstep, and the conditionality ensues as a matter of course. Not a flinch. Not a flicker.

In sum, we have two *H* gates here, which will flip nuclear spins conditionally based on the values the spins encode. These gates are more than the equivalent of two *H* gates in series (see box 18.1); they constitute in effect a conditional *NOT* gate—or "controlled" *NOT* gate, as is the running cliché.

---

*The difference amounts to a few thousand Hertz in a strong magnetic field.

The field of quantum computing is still very young and is limping along with a variety of names for the same things. Not knowing which one will catch on, I used both.

Toward Multi-Qubit Computers

It bodes well for the future of quantum computing that the basic instrumentarium of quantum logic is present in rather ordinary molecules. Even a little one, like bromothiophene, has some dormant computing talent. But think how much more talent might lurk in a long hydrocarbon chain or a really big biomolecule!

There is plenty of material out there waiting to be mined. The two-qubit model we have seen was but the opening wedge, and it's time now to scale up the model and be on the lookout for suitable molecular materials. But what kind of molecules should one go for? What guideposts are there for the would-be quantum-computer engineer? I say engineer, but I also mean Lady Evolution, for she must have faced a similar problem in bygone times—of which more later.

One's gut reaction to the questions is to go for big molecules. The larger the molecule, the more nuclear spins it can hold, and there is no dearth of molecules with multiple spins. Alas, things are really not so straightforward—well, they never are in the quantum sphere. Molecular enlargement spawns its own problems: some of the spins get to be farther apart, and the interactions between the most distant ones eventually become too weak to serve as conditional *NOT* gates; besides, the computer readout, the signal that bears the result of the computing, becomes fuzzy, as it is fished out from a background of randomness and decays exponentially with the number of magnetic nuclei.*

So we can't pin our hopes on molecular size alone and must judge each case on its own merits. Nevertheless, to forage for larger molecules is basically sound, and finding molecules with enough nuclear

---

*The signal gets extracted through averaging from a small preferred set of nuclear spins in thermal equilibrium here (room temperature), a set sharing the same quantum state.

spins should be no problem. Hydrogen atoms are everywhere, and their nuclei (^{1}H) are intrinsically magnetic (see table 18.1). Besides, the natural molecular repertoire may be expanded by pressing carbon into service here. This takes a little tinkering, because the common carbon atom (^{12}C) has no spin. But carbon atoms with an extra neutron imparting a spin can be made and then incorporated into an organic molecule. Gershenfeld and Chuang already led the way in that direction with a chloroform molecule ($CHCl_3$) of this sort. And putting the interplay between its carbon and hydrogen nucleus to use as a logic gate, they successfully ran a quantum search algorithm with it.

Table 18.1. Spin ½ Nuclei and Their Energy Scales Expressed as Resonance Frequency in Mega-Hertz

^{1}H	^{19}F	^{31}P	^{13}C	^{15}N
500 MHz	470 MHz	202 MHz	125 MHz	50 MHz

Molecular enlargement by itself is no panacea, and we'd better hedge our bets until more experimental information is on hand. But what is on hand so far is bright with promise. Since the bromothiophene prototype, the number of qubits in subsequent liquid molecular models has been steadily increasing, culminating in a seven-qubit one built by Chuang and his colleagues. That boasts seven nuclear spins—two carbon (^{13}C) and five fluorine (^{19}F) nuclei (see figure 18.4)—each able to adopt two orientations, spin-up and spin-down. Each can be addressed separately because of adequate spread of the resonance frequencies. As in the prototype, that spread is the key to logic gating; in fact, the molecule was specially synthesized for that. But the proof of the pudding is in the eating: the seven-spin molecule factored the number 15.

This doesn't exactly break the bank. Nevertheless, it is a major technological achievement. It is the first computer to implement a complex quantum algorithm (Shor's algorithm). For information theory, it is nothing less than a milestone: it offers the first glimpse of the material side of that gossamery entity, Information. It took half a century for Information to show its flipside, but it is only right that this had to wait until one could dig for it at the very bottom of physics.

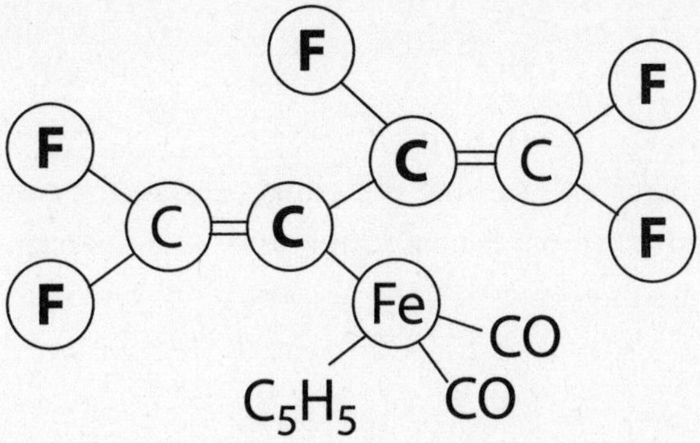

Figure 18.4. A Quantum-computer Molecule with Seven Nuclear Spins. The pertinent carbon (^{13}C) and fluorine (^{19}F) nuclei are in boldface; the other carbons are ^{12}C. The molecule is a synthetic perfluorobutadienyl complex. (After Vandersypen et al. 2001.)

Cory and his colleagues, for their part, made a go of the nuclear spins in solid-state molecules. The coupling between spins here is stronger than in liquid molecules, which makes for faster logic gating—a benefit of a solid lattice. The investigators used crystals of calcium fluoride (a toothpaste ingredient). Each crystal was about half an inch across, offering a billion billion spins for nuclear magnetic resonance programming. Taking advantage of that vast quantum parallelism, the investigators managed to perform quantum simulations involving billions of qubits—an operational scale of information processing unimaginable in a classical digital computer.

Sometimes, brilliance flashes from the brows of men as well as Zeus's.

CHAPTER NINETEEN

Quantum Information Processing and the Brain

Naturally Slow Decoherence

In the foregoing pages I may have hurried things too much. Keen on hoisting the quantum flag, I glossed over why those molecular models were so robust, why their superposed quantum states held up so well under the environmental onslaught. No minor question this—all that innate molecular computer talent would be not worth a damn if the pertinent quantum states were to decohere before the atoms had time to run through their quantum-logic sequences. It was plain cheek when I said the steadfast scientist sometimes gets a reprieve. To expect nature to declare a moratorium on her irreversible processes is to expect the impossible—and here we are up against the most fundamental, most unforgiving of such processes. The best we may hope for is to slow down decoherence enough by insulating the atoms from the perennial environmental electronic and vibrational influences.

And that is precisely how those molecular models came out ahead. They use atomic nuclei that have a small magnetic moment (spin ½; see table 18.1) and couple poorly to other particles. This is the reason, or one reason anyway, that the quantum states in those molecular models are so exceptionally robust. They all have notoriously long nuclear-magnetic relaxation times—up to half a minute—long enough for logic gating to be over and done with.

Felix Bloch, the father of nuclear-magnetic-resonance spectroscopy, already had an inkling of that. In his classic paper on nuclear induction, written decades before there was any knowledge of quantum decoherence (1946), he noted that nuclei with spins ½ have "inconveniently" long relaxation times (the analog of coherence times). But what is inconvenient for the chemist is a boon for the quantum-computer engineer— and that includes the oldest and most experienced one, Lady Evolution.

I'll say more about that later on. First a word about the fundamentals of her endeavor, as seen from the physics bottom.

A View of Evolution from the Quantum Bottom

At the quantum strata of that endeavor, the basic level, the information gets processed very differently from what we see higher up. However, it is not the information itself that is different. Information is information, whether it is carried by atomic nuclei, molecules, or the balls of an abacus. It is always the same gossamery stuff derived from the primordial Information Arrows.

We saw those arrows radiating out from the initial information state, the cosmic singularity, engendering all manner of organizations (figure 2.2). We picked among the arrows—and it was out of sheer self-interest—the one bearing for our little perch in the universe. But all the others are made of the same gossamery stuff. And *gossamery* is the right word. No matter how physicists and information theorists dealt with it, whether by using Boltzmann's statistical mechanics, Shannon's equation, complexity theory, or Planck's energy-frequency relationship, and no matter how exact the quantifications, it always was, and could not escape being, something ethereal.

Well, no longer. The astonishing experiments with quantum-computing molecules dealt with in the preceding chapter give us an almost palpable experience. What is more, they give us a new perspective on the molecular world, *where molecules engage in internal quantum information processing and computing as much as in exchanging with each other macroscopic chunks of information.*

This is a much broader perspective than we ever had before—a view from the physics bottom—showing us that it is the *totality* of the information flows mentioned above, not just the flow between molecules, that propels the Information Arrow and hence the evolution of all organization on Earth. It is a view wherein the evolutions of inanimate and living matter are part of one and the same information landscape.

But perhaps the most satisfying side of that view—if I may interject a personal note—is the realization that biology no longer lacks the symmetries, the perfectly balanced epiphanies, of physics.

Two Variations on an Ancient Biological Quantum Theme

That realization gives us license to phase out the time-honored way of treating biological evolution and physical evolution separately. I know, this will meet with opposition from some card-carrying evolutionists. That can't be helped. It is difficult to argue with those who haven't yet learned to hold their own myths in check. The traditional way may be expedient when dealing with animal species, but it becomes contrived, even untrue, when viewed from the quantum bottom.

Such viewing immediately also cuts the ground from under old misapprehensions about the relevance of quantum physics to biology, sweeping out the last cobwebs that may have lingered in our minds. At long last we have now reached a mark where we may begin to categorize biologically significant quantum-mechanical phenomena and integrate them into the scheme of life. There is hard evidence for them at basic levels of physiology. In keeping with the theme of this book, we have concentrated on evidence concerning the brain. But there is more, and it is evidence of the hardest sort and at the most fundamental level of life: photosynthesis.

Recall that the primary event in photosynthesis is the capture of information from solar photons by pigment molecules of plant and bacteria (figure 2.4). The capture is a spectacular feat, and it is brought about by electrons of these molecules. But no less spectacular is the ulterior transmission of the information by the electrons. That proceeds almost without any losses, and such extreme efficiencies are the signature of coherent quantum waves (chapter 5). The oldest and best-known example in physics is superconductivity, a quantum-mechanical state in which matter loses all electrical resistance. Something akin is happening here.

The evidence comes from experiments Graham Fleming and Robert Blankenship and their colleagues performed on the chlorophyll complex of a green sulfur bacterium (*Chlorobium tepidum*). Using state-of-the-art technique (two-dimensional electronic spectroscopy) providing femtosecond time resolution (10^{-15} seconds), they showed that the initial electron transfer here is by coherent quantum waves.

The coherency is short-lived. But it lasts at least 660 femtoseconds, and in the multidimensional quantum space where things can go forward *and* backward in time, that's enough to sample among many superposed states of quantum waves the one where energy finds the optimal sink to flow into—which is precisely how this little biomolecule achieves the amazing energy efficiency. Or put in terms of information instead of energy, in those 660 femtoseconds of to-and-fro the molecule implements an algorithm enabling it to select, out of the many quantum states hovering about, the one that provides the most efficient information transmission.

In short, the photopigment molecule performs a quantum computation.

We now may also bring that information view to bear on the events happening at the photon-sensing outposts of our brain, the sensory cells in our eyes. There the capture of the photon information takes place at the retinal carbon chain of rhodopsin, and the atomic nuclei hovering about the chain and its α-helix linkage transmit the information as coherent quantum waves down the chain into the rhodopsin molecule (figure 5.2). As in the ancient photosynthesis pigment molecule, each impinging photon triggers a round of quantum computation—the requisite quantum-logic gates and nonlinearity are built into the atomic structure of the molecule. All it takes is the right photon to come along and get things going. From then on the operation runs like clockwork, the molecules enacting an algorithm that gives maximal transmission efficiency.

Rhodopsin is aeons younger than the photosynthesis molecules, but it fully shows its kinship under the information loupe. Its retinal carbon chain is the functional analog of the ancient molecule—a chip off the old block—and the knack of quantum computing runs in its blood.

The computing by the rhodopsin molecule is faster. The entire quantum cabal is over in barely 200 femtoseconds, and that's all the time the molecule has before decoherence sets in. But it is thanks to that brief instant of grace that our brain is apprised of even the slightest glimmer of light in the world outside.

A Final Darwinistic-Physics Inquirendo and a Hypothesis

Perhaps it would be wise to leave it at that. This is as far as the harder evidence goes for physiologically significant quantum-mechanical phenomena in the brain. But there is a message in that bottle, and I cannot resist the temptation to uncork it.

To that end I will take once more the Darwinistic-physics path, and I beg the reader's indulgence. But that path has stood us in good stead on several other occasions when we tried to pry into Evolution's secrets, especially in matters concerning photons and electrons. Recall the answers we got to questions like why white is white or why Evolution would choose the electrons in the first place (chapter 5). Those questions were intertied, but the answer to the second question has a direct bearing on the matter in hand. So let me restate the gist of it:

> Evolution's choice was conditioned by what was available in her earthly space and affordable informationwise. And available were two energy fields—a field of electron quanta and a field of photon quanta—and where and when the quanta matched, there was information to be had for free.

This was an offer the lady couldn't refuse. But it was an opportunity only available on the quantum floor, the ground floor of energy and matter.

And that's the place to be if you are a scrooge, as all sorts of particles collide there as a matter of course, tendering a nonlinear framework for quantum computation.

That is easily said now, though to be candid, scientists were rather slow on the uptake. We wised up to those opportunities only a few years ago. But after that there was no stopping the enterprise, especially when it became feasible to engineer molecules with nuclear spin suitable for quantum computation (e.g., figure 18.4). And what is feasible for today's engineer should have been no less so for the oldest and most consummate molecular engineer, Evolution.

So we are back to our theme song, but now at the top of our voices, as we know for certain that molecules capable of quantum computations came off her assembly lines at least twice: early on in her enterprise, in a simple form serving the function of photosynthesis, and later in more ornate form, serving the function of vision in the brain. Thus it is not farfetched to think that under the relentless pressures for speed of organismic reaction, she would come up with more quantum-computing molecular forms serving other functions in the brain.

One function comes readily to mind: parallel computation. This is the operation conjoining our computational brain, the overarching function of its vast web of neurons—and what could possibly be more suitable for that than quantum computation?

So, bit by bit, the outline of a hypothesis takes shape:

> The neuronal web of our brain can operate in two computational modes: a quantum mode where quantum waves are the substrate, and a macroscopic mode where large packets of ions (the well-known electrical nerve signals) are the substrate. The first mode serves the function of parallel computation and the second that of integration and low-order polynomial computation.

I deliberately put the quantum mode first because it is more information-economical and so likely to have been first on the bioevolutionary scene, well before the two modes became coextensive.

Neuronal Quantum Computing
in Light of the Hypothesis

It takes a while to warm to the notion of a quantum-computing brain—
it's never easy to break the ice in matters concerning the bewildering
quantum world. But in this case it is perhaps even less so, because the
macroscopic electrical signals are so glaringly conspicuous in peripheral nerves, spinal cord, and brain that one is wont to shrug off the
thought of something as esoteric as quantum waves. However, when
those waves are harnessed for operations of logic, they will do wonders,
as we saw in chapter 18.

Let me say a few more words about those wonders. They all come
down to an eerie ability of the waves to be in superposition, that is, of
many waves to coexist in parallel in quantum space. This parallelism
is the mark of quantumness and the essence of quantum computing.
A quantum computer thus is by its very nature a parallel computer—
its whole processing architecture is parallel. And as long as its unit-
computing elements (the quantum waves) stay in superposition, it is
capable of being in many information states at the same time (up to 2^n
states).* Such a computer needn't even have a large number of quantum
bits (n); even a modest quantum computer, say, a 12-qubit model, can
be in 2^{12} states (about 4,000) at once. It therefore doesn't have to go
through the time-consuming, step-by-step procedure of serial computing (like the step-by-step procedure of long multiplication we learned
in primary school) standard digital computers have to go through. One
single computer here can perform thousands of parallel operations at
the same time, operations that would require thousands of standard
digital computers hooked up in parallel.

---

*In general, a string of n quantum bits can exist in any state of the form

$$\Psi = \sum_{x=00...0}^{11...1} c_x |x\rangle$$

Where c_x are complex numbers and the index x ranges over all 2^n classical values of an
n-bit string. Quantum information processing consists of applying a sequence of unitary transformations to the state vector Ψ.

Now, consider our brain. Its web of neurons is endlessly engaged in parallel processing and computing of sensory information. It's computing on a massive scale. You need to look no further than what happens to the information coming from an ordinary object in the world outside. The bits get divided up at the level of our senses and funneled into discrete information channels, to be processed separately and simultaneously: the bits pertaining to shape go to one compartment; those of color, to another; those of texture, to yet another—and so on and on, the bits of smell, taste, and movement all going to their separate brain compartments and undergoing processing in parallel (chapter 14). This way, enormous numbers of bits of external information can be handled in an instant and the results be made available to conscious perception in an instant. And I don't use *enormous* loosely here. These are truly astronomical amounts of information—even the number of bits bound up with a brief snapshot of visual perception, the number in a single activity matrix underlying a conscious state, is astronomical.

Although that assessment was intended for our brain, those astronomical quantities of information are by no means unique to *Homo sapiens*. It is just hubris to think they are. We tend to make too much of a few hundred thousand generations, which is all that separates us from the nearest ape, though we know well that we share most of its DNA. The little we don't share, it is true, can make a big difference in behavior and cognition, but not necessarily in information-processing capacity—and certainly not in basic information-processing structure. Hardware is more difficult to change than software.

Indeed, going down the zoological ladder, we find parallel sensory-information processing everywhere. The amounts of information a cat, a frog, a bird, or even a fly will handle in an instant are vast; the number of bits bound up in a sensory activity matrix may in some cases even be larger than in our brain. The dog obviously has one up on us in smell in this regard,* and just think what mother lodes the massive brain of a whale may bear!

---

*It is not just the sensitivity of individual information channels that makes the dog's olfactory system superior, but the wider variety of the channels. Our system has shriv-

Looking back over the aeons, we are now beginning to get a sense of Evolution's neuronal strategy. This appears from early on to have been geared to parallel information processing—and quantum computing fits that to a T. Such computing may seem esoteric, but all the unfamiliar is wont to do so. And things aren't precisely helped by our laptops. Our daily use of these digital devices makes us think that their computing is easy—easier than quantum computing. But we have it backward. We are overlooking the huge information cost of macroscopic computing—and digitized macroscopic computing as practiced by neurons is in no way less expensive. The hardware alone (the ion-channel proteins, their lipid framework, and control of channel throughput) has a stiff price, not to speak of the price the thermodynamics piper demands.

Yes, we have it backward. In Evolution's dictionary, easy is synonymous with information-economical, and to call digital macroscopic computing easy is as politically incorrect as eating a whale-meat sandwich.

Explanatory Remarks on the Hypothesis

A few additional words about the hypothesis here proposed are in order. The hypothesis bears no relation—and I wish to emphasize this point—to theories positing a quantum-wave entanglement with consciousness, nor to theories about quantum gravity (see chapter 17). In the present hypothesis, quantum computing is biologically important in its own right: it is useful for fast-track computer operations in the sensory brain—or put in stronger developmental terms, it occupies, along with distributed coding and logic depth, the age-old evolutionary niche of saving computational time and effort.

Perhaps just as important as what the hypothesis is, is what it is not. It is not a hypothesis of consciousness. It is a hypothesis limited to the computer operations of the brain, and it bears on consciousness only insofar as those operations are the prolegomena in this mystery of mysteries.

---

eled in the course of evolution; many of our ancestral odor-receptor genes have become inactive. The genes are still there—molecular biologists can read them in our DNA script, but our translational machinery cannot.

Prospects

Before us now lies an exciting prospect: a search for quantum waves at the brain centers, quantum waves involved in functions of parallel computation. It will be a formidable task, and I can speak with feeling—I've been there. Herbert Fröhlich and I were on a quest like that with a more limited scope in the 1960s, a search for quantum waves in cell junctions, the tightly adhering spots of cell membranes. Our focus of interest was then a broader and more elementary biological function: the transmission of information between epithelial cells. That search was guided by the knowledge that quantum coherence is rather robust in dielectric crystals—Fröhlich, a pioneer in the physics of superconductivity, had worked out the theory behind that. And because the cell membrane is a good dielectric (chapter 4), it seemed a good bet that it might carry quantum signals from cell to cell. We probed the junctions of a number of epithelial cell types. Alas, we came up empty-handed. There is a good deal of background noise in living cells, and noise is the perfect spoiler. Even in our best and most promising experiments, resonance in thermal frequency modes reared its ugly head, rumpling our spectroscopic signal readout. Our tools then just weren't good enough.

But the scene looks a lot brighter now. Better tools are on hand—a whole arsenal of them, ranging from infrared spectroscopy, to electron spectroscopy, to laser scattering, to nuclear magnetic resonance—providing much higher signal resolution. Moreover, we now have the experience, gained from actual quantum computing with molecules, to lean on—and nothing beats hands-on experience when it comes to experimental science. That know-how is only a few years old, but it will be invaluable for spotting biological molecules, especially macromolecules, that have natural quantum-logic gates.

Even as I write these lines, such experience is offering a helpful hint regarding macromolecules. It concerns the general noise problem with large molecules in thermal equilibrium, the cruel paradox of seeing one's readout signals fade with the increasing number of nuclear spins in a molecule (chapter 18). As it turns out, the problem can be overcome by polarization. This solution hadn't been known. Quantum-computer engineers had overlooked the polarization factor—but it's unlikely Lady Evolution had.

So, all things considered, there are reasons to be optimistic, though I am reminded of what Chairman Mao said in the 1960s when asked what he thought of the French Revolution: "It's too soon to tell."

Appendix

(1) **Boltzmann's Equation**, as inscribed on his gravestone at the cemetery in Vienna:

$$S = k \log W$$

(2) **Boltzmann's Equation** spelled out to see the formal equivalence with Shannon's Equation of Information:

$$S = -k \sum_i p_i \ln p_i$$

where S is entropy and k Boltzmann's constant, 3.2983×10^{-24} calories / °C

(3) **Shannon's Equation:**

$$I = \sum_i p_i \log_2 p_i$$

Information (I) and entropy (S) have opposite sign but otherwise differ formally only by their scaling factor. An entropy unit equals $-k \ln 2$ bit

(4) **The Cognition Equation:**

$$\Delta I\mathbf{m} = \Delta S\mathbf{e} + \Delta S\mathbf{m}$$

For definition of terms see chapter 2.

(5) **Bennett's Definition of Logical Depth:**

$$\min \{ T(p) : (|p| - |p^*| < s) \wedge (U(p) = x \},$$

the least time required to compute the depth of finite strings $D_s(x)$ by an s-incompressible program, where x is the string and s a significance parameter.

(6) Electrical Analog of a Mechano-Sensory Dendrite (Pacinian Corpuscle)

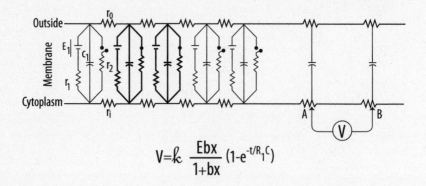

$$V = k \frac{Ebx}{1+bx}(1 - e^{-t/R_1 C})$$

The circuit represents the membrane near the ending of the dendrite during local mechanical stimulation (figure 4.2). The capacitance and transverse resistance are uniformly distributed over the membrane of the cylindrical dendrite; c_1 and r_1 are the corresponding fractional units. (E_1) transmembrane voltage, (r_i) resistance of the cell interior, (r_o) resistance of exterior (very small compared to r_i). The closing of the shunt resistors r_2 represents the opening of channels in the stimulated membrane locale (thick-lined portions of network). The shunting produces a signal (ΔV), which attenuates with distance from the stimulated locale. The equation gives to a first approximation the signal—the generator potential—as measured at some distance from the stimulated locale (between A and B); x is the fraction of membrane channels opened, $1+b$, the ratio between E and the value it drops to in the excited membrane locale, C and R, the lumped c_1s and r_1s (in parallel with those of the excited locale), and k, a constant depending on the location of the measuring electrodes. (After Loewenstein 1959.)

(7) **Mechanical Information Transmission in Pacinian Corpuscle**

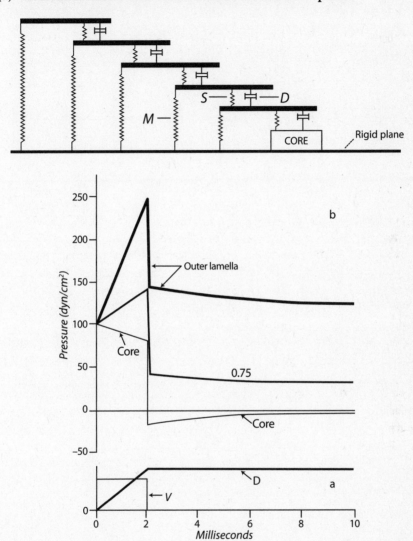

Top: Mechanical analog of the corpuscle. Lamellae are represented by horizontal bars; their compliance, by springs M; the compliance of radial lamellae connection, by weaker springs S; and the fluid resistance, by dashpots D (in parallel with S).

Bottom: Pressures inside the capsule during a compression of 20 thousandths of a millimeter uniformly in 2 thousandths of a second and then maintained. (*a*) time course of displacement (D) and velocity (V) of outermost lamella. (*b*) time course of the computed pressure on the outermost lamella; at 0.75 of capsule radius, and at the capsule center (core). (From Loewenstein and Skalak 1966.)

Recommended Reading

Neurosciences

Adrian, Lord Edgar. 1946. *The Physical Background of Perception.* Oxford: Clarendon Press.

Chalmers, D. J. 1996. *The Conscious Mind.* New York/Oxford: Oxford University Press.

Changeux, J.-P. 1977. *The Neuronal Man: The Biology of Mind.* Princeton, NJ: Princeton University Press.

Churchland, P. S., and Sejnowski, T. J. 1996. *The Computational Brain.* Cambridge, MA: MIT Press.

Damasio, A. 2003. *Looking for Spinoza: Joy, Sorrow and the Feeling Brain.* Orlando, FL: Harcourt.

Dennett, D. 1978. *Brainstorms.* Cambridge, MA: MIT Press.

Dowling, J. E. 1987. *The Retina: An Approachable Part of the Brain.* Cambridge, MA: Harvard University Press.

Hubel, D. H. 1995. *Eye, Brain, and Vision.* New York: Scientific American Library.

Jackendorff, R. 1987. *Consciousness and the Computational Mind.* Cambridge, MA: MIT Press.

Kandel, E. R. 2006. *In Search of Memory.* New York: Norton.

Kandel, E. R., Schwartz, D. N., and Jessell, T. M. 2000. *Principles of Neural Science.* New York: McGraw-Hill.

Koch, C. 2004. *The Quest for Consciousness: A Neurobiological Approach.* Englewood, CO: Roberts & Co.

Nicholls, J. G., Martin, A. R., Wallace, B. G., and Fuchs, P. A. 2000. *From Neuron to Brain.* Sunderland, MA: Sinauer Publishers.

Mountcastle, V. B. 1957. *Perceptual Neuroscience.* Cambridge, MA: Harvard University Press.

Sacks, O. 2011. *The Mind's Eye.* New York: Alfred Knopf.

Searle, J. R. 1992. *The Rediscovery of the Mind.* Cambridge, MA: MIT Press.

Thompson, E. 2007. *Mind in Life.* Cambridge, MA: Harvard University Press.

Physics and Information Theory

Barrow, J. D., Davies, P., and Harper, C., eds. 2005. *Science and Ultimate Reality: Quantum Theory, Cosmology, and Complexity.* London: Cambridge University Press.

Chaitin, G. J. 1999. *The Unknowable.* Singapore: Springer-Verlag.

Davies, P. 1995. *About Time.* New York: Simon & Schuster.

Deutsch, D. 1997. *The Fabric of Reality.* London: Penguin Books.

Dyson, F. J. 2005. Thought experiments in honor of John Archibald Wheeler. In *Science and Ultimate Reality: Quantum Theory, Cosmology, and Complexity,* edited by Barrow, J. D., Davies, P., and Harper, C. London: Cambridge University Press.

Feynman, R. P. 1995. *Six Easy Pieces.* Cambridge, MA: Perseus Books.

Feynman, R. P., Leighton, R. V., and Sands, M. L. 1989. *The Feynman Lectures of Physics.* Reading, MA: Addison-Wesley.

Gleick, J. 2011. *The Information: A History, a Theory, a Flood.* New York: Pantheon Books.

Greene, B. 2004. *The Fabric of the Cosmos.* New York: Alfred Knopf.

Hofstadter, D. 2007. *I Am a Strange Loop.* New York: Basic Books.

Johnson-Laird, P. N. 1988. *The Computer and the Mind.* Cambridge, MA: Harvard University Press.

Lloyd, S. 2006. *Programming the Universe.* New York: Alfred Knopf.

Nielsen, M. A., and Chuang, J. L. 2005. *Quantum Computation and Quantum Information.* Cambridge, UK: Cambridge University Press.

Smolin, L. 1997. *The Life of the Cosmos.* New York: Oxford University Press.

Wüthrich, K. 1995. *NMR in Structural Biology.* Singapore: World Scientific.

References

Preface

Herbert, S. 2006. *Charles Darwin, Geologist.* Ithaca, NY: Cornell University Press. (Darwin's notebook referred to in the text is dated 1838.)

1. Our Sense of Time: Time's Arrow

Our Awareness of Time/The Stream of Consciousness

Dennett, D. C. 1991. *Consciousness Explained.* Boston: Little, Brown.
James, W. 1997. Consciousness. In *The Nature of Consciousness,* edited by Block, N., Flanagan, O., and Güzeldere, G. Cambridge, MA: MIT Press.
St. Augustine of Hippo. *Confessions.* Book XI, xiv (17). New York: Oxford University Press, 1991, p. 230.

Inner Time versus Physics Time

Barbour, J. 1999. *The End of Time.* New York: Oxford University Press.
Bondi, Hermann. 1952. Relativity and indeterminacy. *Nature* 169:660.
Carroll, Lewis. (1865) 1982. *Through the Looking-Glass and What Alice Found There.* New York: Avenel Books.
Davies, P. C. W. 1974. *The Physics of Time Asymmetry.* Berkeley: University of California Press.
Davies, P. 1995. *About Time: Einstein's Unfinished Revolution.* New York: Simon & Schuster.
Einstein, A. 1911. Bemerkungen zu den P. Hertzschen Arbeiten "Über die mechanischen Grundlagen der Thermodynamik". *Annalen der Physik* 34:175–176.
North, J. D. 1972. The time coordinate in Einstein's restricted theory of relativity. In *The Study of Time,* edited by Frazer, J. T., Haber, F. C., and Müller, G. H. Berlin/New York: Springer-Verlag.

Penrose, R. 1991. *The Emperor's New Mind.* London: Penguin Press.

Penrose, R. 1994. *Shadows of the Mind: A Search for the Missing Science of Consciousness.* New York: Oxford University Press.

Smolin, L. 1997. *The Life of the Cosmos.* New York: Oxford University Press.

Taylor, E. F., and Wheeler, J. A. 1992. *Spacetime Physics: Introduction to Special Relativity* (chapter 3). San Francisco: W. H. Freeman.

Wheeler, J. 1957. On the nature of quantum geometrodynamics. *Annals of Physics* 2:604.

Zeh, H. D. 1989. *The Physical Basis of the Direction of Time.* Berlin: Springer-Verlag.

The Way the Cookie Crumbles/How to Predict the Future from the Past/Why the Cookie Crumbles/Can Time Go Backward?

Boltzmann, L. 1877. Über die Beziehung eines allgemeinen mechanischen Satzes zum zweiten Hauptsatze der Wärmetheorie. *Sitzungsberichte der Kaiserlichen Akademie der Wissenschaften Wien, Mathematisch-Naturwissenschaftliche Klasse* 75:67–73.

Boltzmann, L. 1909. *Wissenschaftliche Abhandlungen.* Leipzig: F. Hasenöhrl.

Borges, J. L. 1998. Shakespeare's memory. *New Yorker,* April 13, 667.

Carroll, Lewis. (1865) 1982. "Alice in Wonderland." In *Through the Looking-Glass and What Alice Found There.* New York: Avenel Books.

Halliwell, J. J. 1994. Quantum cosmology and time asymmetry. In *The Physical Origins of Time Asymmetry,* edited by Halliwell, J. J., Perez-Mercader, J., and Zurek, W. H. Cambridge, UK: Cambridge University Press.

Halliwell, J. J., Perez-Mercader, J., and Zurek, W. H. 1994. *The Physical Origins of Time Asymmetry.* Cambridge, UK: Cambridge University Press.

Hawking, S. W. 1988. *A Brief History of Time.* New York/Toronto: Bantam Books.

Hawking, S. W. 1992. Evaporation of two-dimensional black holes. *Physical Review D* 46:603.

Hertz, P. 1910. Über die mechanischen Grundlagen der Thermodynamik. *Annalen der Physik* 33:225–274 and 537–552. (These articles point up the pitfalls besetting any proof of the Second Law and include a historically interesting entry, a rebuke to the young Einstein, who was unaware of them [1902].)

Newton, I. (1687) 1999. *Philosophiae Naturalis Principia Mathematica.* Translated from the 3rd rev. ed. (1726) by Bernard Cohen and Anne Whitman. Berkeley: University of California Press.

Penrose, R. 1979. Singularities and time asymmetries. In *General Relativity: An Einstein Centenary Survey,* edited by Hawking, S. W., and Israel, W. Cambridge, UK: Cambridge University Press.

Price H. 1996. *Time's Arrow and Archimedes' Point.* New York/Oxford: Oxford University Press.

Thorne, K. S. 1994. *Black Holes and Time Warps.* New York/London: Norton.

Zureck, W. H. 1982. "Environment-induced superselection rules. *Physical Review* D 26:1862.

Time and Our Perception of Reality

Einstein, A., and Besso, M. (1903–1955) 1972. *Correspondence.* Edited by Piérre Speziali. Paris: Hermann. (Einstein's remark quoted in the text is from his letter to Vero and Bice Besso, March 21, 1955.)

Fraser, J. T., and Lawrence, N., eds. 1975. *The Study of Time II.* Berlin: Springer-Verlag.

2. Information Arrows

Retrieving the Past

Hawking, S. W. 1988. *A Brief History of Time.* New York: Bantam Books.

Hawking, S. W., and Ellis, G. F. R. 1977. *The Large Scale Structures of Space Time.* New York: Cambridge University Press.

Hogan, C. J. 1998. *The Little Book of the Big Bang.* New York: Springer-Verlag.

Maran, S. P., ed. 1992. *The Astronomy and Astrophysics Encyclopedia.* New York: Van Nostrand and Reinhold.

Rees, M. 1999a. Exploring our universe and others. *Scientific American* 289:78–83.

Rees, M. 1999b. *Just Six Numbers.* New York: Basic Books.

Weinberg, S. 1977. *The First Three Minutes.* New York: Basic Books.

Weinberg, S. 1985. Origins. *Science* 230:15–18.

The Arrows of Time and Information

Boltzmann, L. 1895a. On certain questions of the theory of gases. *Nature* 51:413–451.

Boltzmann, L. 1895b. On the minimum theorem in the theory of gases. *Nature* 52:221.

The Products of Information Arrows: An Overview

Burns, J. O. 1986. Very large structures in the universe. *Scientific American* 254, no. 7:38–47.

Fabian, A. C., ed. 1992. *Clusters and Superclusters of Galaxies.* New York: Kluwer Academic Publishers.

Henry, P. J., Briel U. G., and Böhringer H. 1998. The evolution of galaxies. *Scientific American* 279 (December):52–57.

*The Strange Circles at Coordinates 0,0,0/Molecular Demons/
Wresting Information from Entropy: The Quintessence of Cognition*

Landauer, R. 1961. Irreversibility and heat generation in the computing process. *IBM Journal of Research and Development* 5:183–191.

Loewenstein, W. R. 2000. *The Touchstone of Life: Molecular Information, Cell Communication, and the Foundations of Life.* New York: Oxford University Press; London: Penguin Books.

Maxwell, J. C. (1871) 1880. *Theory of Heat.* 6th ed. New York: D. Appleton.

Shannon, C. E. 1949. Communication in the presence of noise. *Proceedings of the J.R.E.* 37:10.

Who Shoulders Life's Arrow?

Loewenstein, W. R. 2000. *The Touchstone of Life: Molecular Information, Cell Communication, and the Foundations of Life.* New York: Oxford University Press; London: Penguin Books.

3. The Second Coming

The Demons for Fast Information Transmission

Almers, W., and McCleskey, W. 1984. Non-selective conductance in calcium channels of frog muscle: calcium selectivity in a single-file pore. *Journal of Physiology* (London) 353:585–608.

Clapham, D. E. 1999. Unlocking family secrets: K^+ channel transmembrane domains. *Cell* 97:547–550.

Doyle, D. A., Cabral, J. M., Pfuetzner, R. A., Kuo, A. Gulbis, J., Cohen, S., Chait, B. T., and MacKinnon, R. 1998. The structure of the potassium channel: molecular basis of potassium conduction and selectivity. *Science* 280:69–77.

Gulbis, J. M., Zhou, M., Mann, S., and MacKinnon, R. 2000. Structure of the cytoplasmic β subunit-T_1 assembly of voltage-dependent K^+ channels. *Science* 289:123–127.

Hille, B. 1992. *Ionic Channels of Excitable Membranes.* Sunderland, MA: Sinauer Associates.

Hodgkin, A. L., and Keynes, R. D. 1955. The potassium permeability of a giant nerve fibre. *Journal of Physiology* (London) 128:61–88.

Latorre, R., and Miller, C. 1983. Conduction and selectivity in potassium channels. *Journal of Membrane Biology* 71:11–30.

Miller, C. 1999. Ionic hopping defended. *Journal of General Physiology* 113:783–787.

Neyton, J., and Miller, C. 1988. Discrete Ba^{2+} block as a probe of ion occupancy and pore structure in the high-conductance Ca^{2+}-activated K^+ channel. *Journal of General Physiology* 92:569–586.

Nonner, W., Chen, D. P., and Eisenberg, R. 1999. Perspective: Progress and prospects in permeation. *Journal of General Physiology* 113:773–782.

Parsegian, V. A. 1975. Ion-membrane interactions as structural forces. *Annals of the New York Academy of Science* 262:161–171.

Unwin, N. 1989. The structure of ion channels in membranes of excitable cells. *Neuron* 3:665–676.

Yellen, G. 1999. The bacterial K^+ channel structure and its implications for neuronal channels. *Current Opinion in Neurobiology* 9:267–273.

How the Demon Tandems Censor Incoming Information

Loewenstein, W. R. 1965. Facets of a transducer process. *Cold Spring Harbor Symposia on Quantitative Biology* 30:29–43.

4. The Sensors

The Sensory Transducer Unit

Baylor, D. A. 1992. Transduction in retinal photoreceptor cells. *In Sensory Transduction,* edited by Corey, D. P., and Roper, S. D., 151–174. New York: Rockefeller University Press.

Loewenstein, W. R. 1960. Biological transducers. *Scientific American* 203:99–109.

How Electrical Sensory Signals Are Generated

Loewenstein, W. R. 1959. The generation of electrical activity in a nerve ending. *Annals of the New York Academy of Science* 81:367–387.

5. Quantum Sensing

The Quantum World

Bennet, C. H., and Shor, P. W. 1998. Quantum information theory. *IEEE Transactions on Information Theory* 44:2724–2742.

Betrametti, E., and van Fraassen, B., eds. 1981. *Current Issues in Quantum Logic.* New York: Plenum Press.

Boole, G. 1854. *An Investigation of the Laws of Thought.* New York: Dover.

Deutsch, D. 1985. Quantum theory: The Church-Turing principle and the universal quantum computer. *Proceedings of the Royal Society of London, Series A* 400:97–117.

Haack, S. 1974. *Deviant Logic: Some Philosophical Issues.* London: Cambridge University Press.

Hughes, R. I. G. 1981. Quantum logic. *Scientific American* 245:202–213.

Jammer, M. 1974. *The Philosophy of Quantum Mechanics.* New York: John Wiley & Sons.

Zurek, W. H. 1991. Decoherence and the transition from quantum to classical. *Physics Today* 44:36–44.

Our Windows to the Quantum World

Mathies, R. A., and Lugtenburg, I. 2000. The primary photoreaction of rhodopsin. In *Handbook of Biological Physics.* Vol. 3. Edited by Stavenga, D. G., DeGrip, W. J., and Pigh, E. N., Jr. Amsterdam: Elsevier.

Wald, G. 1968. The molecular basis of visual excitation. *Nature* 219:800–807.

Wang, Q., Schoenlein, R. W., Peteanu, L. A., Mathies, R. A., and Shank, C. V. 1994. Vibrationally coherent photochemistry in the femtosecond primary event of vision. *Science* 266: 422–424.

Why We See the Rainbow

Dalton J. 1798. Extraordinary facts relating to the vision of colours. *Memoirs of the Literary Philosophical Society of Manchester* 5:28–45.

Helmholtz, H. von. (1867) 1962. *Treatise on Physiological Optics.* Translated from the 3rd ed. New York: Dover Publications.

Kochendoerfer, G. G., Lin, S. W., Sakmar, T. P., and Mathies, R. A. 1999. How visual pigments are tuned. *Trends in Biochemical Science* 24:300–305.

Maxwell, J. C. 1855. Experiments on colour, as perceived by the eye, with remarks on colour-blindness. *Transactions of the Royal Society, Edinborough* 21:275–298.

Nathan, I. 1985. The genes of color vision. *Scientific American* 260:42–49.

Young, T. 1802. On the theory of light and colours. *Proceedings of the Royal Society London.* A 92:12–48.

Zhukovsky, E. A., Robinson, P. R., and Oprian, D. D. 1991. Transducin activation by rhodopsin without a covalent bond to the 11-cis-retinal chromophore. *Science* 251: 558–560.

Quantum Particles That Don't Cut the Mustard

Longair, A. 1996. The new astrophysics. In *The New Physics,* edited by Davies, P., 94–208, Cambridge, UK: Cambridge University Press.

Maruyama, H. 2000. The origin of neutrino mass. *Physics World* 15:35–39.

Pagels, H. R. 1990. *The Cosmic Code: Quantum Physics as the Language of Nature.* New York: Bantam Books.

Updike, J. 1988. *Telephone Poles and Other Poems.* New York: Knopf. (The poem originally appeared in *The New Yorker* under the title "Cosmic Gall.")

6. Quantum into Molecular Information

Boosting the Quantum

Baylor, D. A., Lamb, T. D., and Yau, K.-W. 1979. Responses of retinal rods to single photons. *Journal of Physiology (London)* 288:613–634.

Baylor, D. A. 1987. Photoreceptor signals and vision: Proctor lecture. *Investigaive Ophthalmology & Visual Science* 28:34–49.

Bourne, H. R., Sanders, P. A., and McCormick, F. 1991. The GTPase superfamily: Conserved structure and molecular mechanisms. *Nature* 349:117–127.

Gilman, A. G. 1987. G proteins: Transducers of receptor-generated signals. *Annual Review of Biochemistry* 56:615–649.

Hecht, S., Schlaer, S., and Pirenne, M. H. 1942. Energy, quantum and vision. *Journal of General Physiology* 25:819–840.

Pugh, E. N., Jr., and Lamb, T. D. 1993. Amplification and kinetics of the activation steps in phototransduction. *Biochimica et Biophysica Acta* 1141:111–149.

A Consummate Sleight of Hand/The Ubiquitous Membrane Demon

Bourne, H. R., Sanders, P. A., and McCormick, F. 1991. The GTPase superfamily: Conserved structure and molecular mechanisms. *Nature* 349:117–127.

Gilman, A. G. 1987. G proteins: Transducers of receptor-generated signals. *Annual Review of Biochemistry* 56:615–649.

Loewenstein, W. R. 2000. *The Touchstone of Life: Molecular Information, Cell Communication, and the Foundations of Life.* New York: Oxford University Press.

Stryer, L. 1986. The cyclic nucleotide cascade of vision. *Annual Review of Neuroscience* 9:87–119.

7. Molecular Sensing

A Direct Line from Nose to Cortex

Firestein, S., Zufall, F., and Shepherd, G. M. 1991. Single odor-sensitive channels in olfactory receptor neurons are also gated by cyclic nucleotides. *Journal of Neuroscience* 11:3565–3572.

Hildebrand, J. G., and Shepherd, G. M. 1997. Mechanisms of olfactory discrimination: Converging evidence for common principles across phyla. *Annual Review of Neuroscience* 20:595–631.

Ottoson, D. 1971. Olfaction. In *Handbook of Sensory Physiology*. Berlin: Springer-Verlag.

Pace, U., Hansky, E., Salomon, Y., and Lancet, D. 1985. Odorant-sensitive adenylate cyclase may mediate olfactory reception. *Nature* 316:255–258.

A Thousand Information Channels of Smell

Axel, R. 1995. The molecular logic of smell. *Scientific American* 273:154–159.

Buck, L., and Axel, R. 1991. A novel multigene family may encode odorant receptors: A molecular basis for odor recognition. *Cell* 65:175–187.

O'Leary, D. D. M., Yates, P., and McLaughlin, T. 1999. Mapping sights and smells in the brain: Distinct mechanisms to achieve a common goal. *Cell* 96:255–269.

Zou, Z., Horowitz, L. F., Montmayeur, J. P., Snapper, S., and Buck, L. B. 2001. Genetic tracing reveals a stereotyped sensory map in the olfactory cortex. *Nature* 414:173–179.

Mapping, Coding, and Synonymity

Einstein, A., 1921. The mot paraphrased in the text is: "Raffiniert ist der Herrgott, aber boshaft ist er nicht" ("The Lord is subtle but he is not mean")—a remark made by Einstein to Oscar Veblen, mathematics professor at Princeton, during one of Einstein's early visits to Princeton in 1921, upon hearing that an experiment by Dayton Miller of Cleveland might contradict Einstein's theory of gravity.

Loewenstein, W. R. 2000. *The Touchstone of Life: Molecular Information, Cell Communication, and the Foundations of Life*. New York: Oxford University Press.

Molecular Sensory Synonymity

Araneda, R. C., Kini, A. D., and Firestein, S. 2000. The molecular receptive range of an odorant receptor. *Nature Neuroscience* 3:1248–1255.

Singer, M. S. 2000. Analysis of the molecular basis for octanal interactions in the expressed rat 17 olfactory receptor. *Chemical Senses* 25:155–165.

8. Electronic Transmission of Biological Information

Evolution's Favorite Leptons/Electronic Information Transmission: A Development Stumped

Boyer, P. D., Chance, B., Ernster, L., Mitchell, P., Racker, E., and Slater, E. C. 1977. Oxidative phosphorylation and phosphophosphorylation. *Annual Review of Biochemistry* 46:955–1026.

Margoliash, E., and Schejter, A. 1967. Cytochrome c. *Advances in Protein Chemistry* 21:113–282.

Mitchell, P. 1979. Keilin's respiratory chain concept and its chemi-osmotic consequences. *Science* 206:1148–1159.

Salemme, F. R. 1977. Structure and function of cytochromes c. *Annual Review of Biochemistry* 46:299–329.

Salemme, F. R., Freer, S. T., Xuong, N. H., Alden, R. A., and Kraut, J. 1973. The structure of oxidized cytochrome c_2 of *Rhodospirillum rubrum. Journal of Biological Chemistry* 248:3910–3921.

Swanson, R., Trus, B. L., Mandel, N., Mandel, G., Kallas, O. B., and Dickerson, R. E. 1977. Tuna cytochrome c at 2.0 Å resolution. I. Ferricytochrome structure analysis. *Journal of Biological Chemistry* 252:759–775.

Two Old Batteries

Davy, H. 1839. Memoirs of his life. In *The Collected Works of Sir Humphry Davy,* Vol. I, edited by Davy, J. London: Smith, Elder and Company.

Thompson, J. J. 1897. Cathode rays. *Philosophical Magazine* 44:293.

Volta, A. 1800. On the electricity excited by the mere contact of conducting substances of different kinds. (in French) *Philosophical Transactions of the Royal Society, London.* 90:403–431.

9. The Random Generators of Biomolecular Complexity

Genuine Transmogrification

Bennett, G. H. 2003. How to define complexity in physics and why. In *From Complexity to Life,* edited by Gregersen, N. H. New York: Oxford University Press.

Chaitin, G. J. 1987. *Algorithmic Information Theory.* Cambridge: Cambridge University Press.

Chothia, C. 1992. One thousand families for the molecular biologists. *Nature* 357:543–544.

Shannon, C. E., and Weaver, W. 1959. *The Mathematical Theory of Communication.* Urbana: University of Illinois Press.

Wilson, A. C., Ochman, H., and Prager, E. M. 1987. Molecular time scale for evolution. *Trends in Genetics* 3:241–247.

Zurek, W. H. 1989. Algorithmic randomness and physical entropy. *Physical Review A* 40:4731–4751.

A Quantum Random Generator of Molecular Form

Haseltine, N. 1983. Ultraviolet light repair and mutagenesis revisited. *Cell* 33:13–17.

Muller, H. J. 1927. Artificial transmutation of the gene. *Science* 46:84–87.

Stadtler, L. J. 1928. Mutations in barley induced by X-rays and radium. *Science* 110:543–548.

The Random Generator and Our Genetic Heritage

Chinwalla, A. T., et al. 2002. Mouse Genome Sequence Consortium: Initial sequencing and comparative analysis of the human genome. *Nature* 420:520–562.

Drake, I. W. 1969. Comparative rates of spontaneous mutation. *Nature* 221:1132.

Loewenstein, W. R. 2000. *The Touchstone of Life: Molecular Information, Cell Communication, and the Foundations of Life.* New York: Oxford University Press.

O'Brien, G. J., et al. 1999. The promise of comparative genomics in mammalia. *Science* 286:458–481.

An Algorithm Is No Substitute for a Demon

Culberson, J. C. 1998. On the futility of blind search: An algorithmic view. *Evolutionary Computation* 6:2.

Dembski, W. A. 2003. Can evolutionary algorithms generate specified complexity. In *From Complexity to Life,* edited by Gregersen, N. H. New York: Oxford University Press.

The Second Generator of Biomolecular Form

Patthy, L. 1991. Modular exchange principles in proteins. *Current Opinion in Structural Biology* 1:351–361.

10. The Ascent of the Digital Demons

Quantum Electron Tunneling

Hopfield, J. J. 1977. Photo-induced charge transfer: A critical test of the mechanism and range of biological electron transfer processes. *Biophysical Journal* 18:311–321.

Potasek, M. J., and Hopfield, J. J. 1977. Experimental test of vibronically coupled tunneling description of biological electron transfer. *Proceedings of the National Academy of Sciences USA* 74:229–233.

Do Plants Have Digital Demons, Too?

Darwin, C. (1863–1866) 2003. *The Correspondence of Charles Darwin.* Vols. XI, XII, XIII. Edited by F. Burkhardt. Cambridge: Cambridge University Press.

Darwin, C. 1875. *Insectivorous Plants.* London: John Murray.

Forterre, Y., Skotheim, J. M., Dumais, J., and Mahadevan, L. 2005. How the Venus flytrap snaps. *Nature* 433:421–425.

Juniper, B. E., Robins, R. J., and Joel, D. M. 1989. *The Carnivorous Plants.* London: Academic Press.

Stuhlman, O., and Darden, B. 1950. The action potentials obtained from Venus's flytrap. *Science* 111:491–492.

Volkov, A. G., Adesino, T., Markin, V. S., and Jovanov, E. 2008. Kinetics and mechanism of *Dionaea muscipula* trap closing. *Plant Physiology* 146:694–702.

11. The Second Information Arrow and Its Astonishing Dénouement: Consciousness

The Structure of Time

Boltzmann, L. 1895. On certain questions of the theory of gases. *Nature* 51:413–415.

Boltzmann, L. (1896/1898) 1964. Vorlesungen über Gastheorie. Leipzig: J. A. Barth. In English translation: *Lectures in Gas Theory*, 74, 446–448. Berkeley: University of California Press; London: Cambridge University Press.

Broda, E. 1983. *Ludwig Boltzmann: Man, Physicist, Philosopher.* English translation by Gray, L. Woodbridge, CT: Ox Bow Press.

12. How to Represent the World

The Universal Turing Machine

Hofstadter, D. R. 1980. *Gödel, Escher, Bach: An Eternal Braid.* New York: Vintage Books, Random House.

Trakhtenbrot, B. A. 1963. *Algorithms and Automatic Computing Machines.* Boston: D.C. Heath & Co.

Turing, A. 1936. On computable numbers, with an application to the Entscheidungsproblem. *Proceedings of the London Mathematical Society* (2nd Series) 42:230–265.

Turing, A. 1937. On computable numbers, with an application to the Entscheidungsproblem: A correction. *Proceedings of the London Mathematical Society* 43:544–546.

Rendering the World by Computer

Deutsch, D. 1985. Quantum theory, the Church-Turing principle and the universal quantum computer. *Proceedings of the Royal Society of London* 400:97–117.

Deutsch, D. 1997. *The Fabric of Reality.* London: Penguin Books.

Milburn, G. 1998. *The Feynman Processor.* Cambridge, MA: Perseus Books.

Our Biased World Picture

Barlow, H. B., Hill, R. M., and Levick, W. R. 1964. Retinal ganglion cells responding selectively to direction and speed of image motion in the rabbit. *Journal of Physiology* 173:377–407.

Hartline, H. K. (1967) 1972. Visual receptors and retinal interaction. In *Nobel Lectures,* 269–288. Amsterdam: Elsevier.

Hartline, H. K., and Ratliff, F. 1957. Inhibitory interaction of receptor units in the eye of *Limulus. Journal of General Physiology* 40:357–376.

Kuffler, S. W. 1983. Discharge patterns and functional organization of the mammalian retina. *Journal of Neurophysiology* 16:37–68.

Loewenstein, W. R. 1965. Facets of a transducer process. *Cold Spring Harbor Symposia in Quantitative Biology* 30:29–43.

Loewenstein, W. R, and Mendelson, M. 1965. Components of receptor adaptation in a Pacinian corpuscle. *Journal of Physiology* 177:377–397.

Loewenstein, W. R., and Skalak, R. 1966. Mechanical transmission in a Pacinian corpuscle. An analysis and a theory. *Journal of Physiology* 182:346–378.

Computing by Neurons

Bennett, C. H. 1982. The thermodynamics of computation. *International Journal of Theoretical Physics* 21:905–940.

Bennett, C. H. 1987. Demons, engines and the second law. *Scientific American* 257:108–116.

Churchland, P. S., and Sejnowski, T. J. 1992. *The Computational Brain.* Cambridge, MA: MIT Press.

Hatsopoulus, N., Gabbiani, F., and Laurent, G. 1995. Elementary computation of object approach by a wide field visual neuron. *Science* 270:1000–1003.

Koch, C. 1999. *Biophysics of Computation: Information Processing in Single Neurons.* New York: Oxford University Press.

Koch, C., Poggio, T., and Torre, V. 1982. Retinal ganglion cells: A functional interpretation of dendritic morphology. *Philosophical Transactions of the Royal Society B* 298:227–264.

Loewenstein, W. R. 2000. *The Touchstone of Life: Molecular Information, Cell Communication, and the Foundations of Life.* New York: Oxford University Press.

Rall, W. 1962. Branching dendritic trees and motoneuron resistivity. *Experimental Neurology* 1:491–527.

Rall, W., Burke, R. E., Holmes, W. R., Jack, J. J. B., Relman, S. J., and Segev, I. 1992. Matching dendritic neuron models to experimental data. *Physiological Review* 72:8159–8186.

Flying the Coop of Our Senses

Blackett, P., and Occhialini, G. P. S. 1933. Some photographs of the tracks of penetrating radiation. *Proceedings of the Royal Society (London) Series A* 139:699–726.

Diamond, J. 1993. *The Third Chimpanzee.* New York: HarperCollins.

Dirac, P. A. M. 1928. The quantum theory of the electron. *Proceedings of the Royal Society (London) Series A*117:610–612.

Huxley, H. 2003. The Cavendish Laboratory. *Physics World* (March):29–35.

Kant, I. 1781. *Kritik der reinen Vernunft.* Leipzig: J. F. Hartknoch.

Kaufmann, W. 1902. Die electromagnetische Masse des Elektrons. *Physikalische Zeitschrift* 4:54–57.

Mendel, J.G. 1866. *Versuche über Pflanzenhybriden.* New York: Hafner.

Roux, W. 1883. *Über die Zeit der Bestimmung der Hauptrichtungen des Froschembryos. Eine biologische Untersuchung von Dr. Wilhelm Roux,* 1–29. Reprinted by W. Engelmann, Leipzig.

Sibley, C. G., and Comstock, Alquist, J. E. 1990. DNA hybridization evidence of hominoid phylogeny: A reanalysis of the data. *Journal of Molecular Evolution* 30:202–236.

Thompson, J. J. 1897. Cathode rays. *Philosophical Magazine* 44:293.

Tobias, P. V. 1995. The brain of the first hominids. In *Origins of the Human Brain,* edited by Changeux, J.-P., and Chavaillon, J. Oxford: Oxford University Press.

Watson, J. D., and Crick, F. H. C. 1953. Genetical implications of the structure of deoxyribonucleic acid. *Nature* 171:964–967.

Weinberg, S. 1992. *Dreams of a Final Theory.* New York: Pantheon.

13. Expanded Reality

A Fine Bouquet

Einstein, A. 1956. Autobiographische Skizze. In *Helle Zeit—Dunkle Zeit: In Memoriam Albert Einstein,* edited by Seelig, C. Zurich: Europa.

Főising, A. 1998. *Albert Einstein.* London and New York: Penguin Books.

The Collected Papers of Albert Einstein. 1987. Edited by Johan Stachel et al. Princeton, NJ: Princeton University Press.

Mathematics and Reality

Greene, B. 1999. *The Elegant Universe.* New York: Vintage Books, Random House.

Greene, B. 2004. *The Fabric of the Cosmos: Space, Time, and the Texture of Reality*. New York: Alfred Knopf.

Holland, J. H., Holyoak, K. J., Nisbett, R., and Thagard, P.R., 1989. *Induction Process of Inference, Learning and Discovery*. Cambridge, MA: MIT Press.

Penrose, R. 2005. *The Road to Reality*. New York: Alfred Knopf.

Plato. 1901. *The Republic*. Translated into English. New York: The Colonial Press.

Plato. 1930. *Phaedo*. Translated into English. Waltham Saint Lawrence, Berkshire: Golden Cockerel Press.

The Neuron Circuitry of Language

Broca, P. 1863. Localisations des functions cérébrales. Siège de la faculté du language articulé. *Bulletin de la Societé d'Anthropologie* 4:200–208.

Calvin, W. H. 1991. *The Ascent of Mind: Ice Age Climate and the Evolution of Intelligence*. New York: Bantam Books.

Changeux, J.-P., and Connes, A. 1999. *Conversations on Mind, Matter and Mathematics*. Princeton, NJ: Princeton University Press.

Chomsky, N. 1975. *Reflections on Language*. New York: Pantheon.

Fuster, J. M. 1980. *The Prefrontal Cortex*. New York: Raven Press.

Pinker, S. 1994. *The Language Instinct*. New York: HarperCollins.

Wernicke, C. 1874. *Der aphasische Symptomenkomplex: Eine psychologische Studie auf anatomischer Basis*. Breslau: Max Cohn & Weigert.

The Reluctant Sensory Brain

Einstein, A. 1905. Zur Elektrodynamik bewegter Körper. *Annalen der Physik* 17:891–921.

Einstein, A. 1905. Ist die Trägheit eines Körpers von seinem Energieinhalt abhängig? *Annalen der Physik* 18:639–641.

The Limits of Knowledge

Chaitin, G. J. 1998. *The Limits of Mathematics*. New York and Heidelberg; Springer-Verlag.

Chaitin, G. J. 1999. *The Unknowable*. Singapore: Springer-Verlag.

Church, A. 1936. An unsolvable problem of elementary number theory. *American Journal of Mathematics* 58:345–363.

Gödel, K. 1931. On formally undecidable propositions of Principia Mathematica and related systems I. *Monatsschrift für Mathematik und Physik* 38:131–198.

Prigogine, I. 1980. *From Being to Becoming: Time and Complexity in Physical Sciences*. San Francisco: Freeman.

Turing, A. M. 1936/1937. On computable numbers, with an application to the *entscheidungsproblem. Proceedings of the London Mathematical Society Series 2* 42:230–265 and 43:544–546.

A Note about Reality

Carroll, Lewis. (1871) 1982. *Through the Looking Glass and What Alice Found There*, Chapter VII. New York: Avenel Books.

14. Information Processing in the Brain

Cell Organization in the Brain

Braak, H. 1976. On the striate area of the human isocortex. *Journal of Comparative Neurology* 166:341–364.

Hubel, D. H. 1988. *Eye, Brain, and Vision*. Scientific American Library. New York: W. H. Freeman & Co.

Koch, C. 2004. *The Quest for Consciousness: A Neurobiological Approach*. Englewood, CO: Roberts & Co.

Lorente de Nó, R. 1938. Cerebral cortex: Architecture, intracortical connections, motor projections. In *Physiology of the Nervous System*, edited by Fulton, F. J., 291–339. New York: Oxford University Press.

Mountcastle, V. B. 1957. Modality and topographic properties of single neurons of the cat's somatic sensory cortex. *Journal of Neurophysiology* 20:408–434.

Ramón y Cajal, S. 1911. *Histologie du Système Nerveux de l'Homme et de Vertébrés*. Paris: Maloine.

Cortical Information-Processing Units/ Cortical Cell Topography and Worldview

Adrian, Lord Edgar. 1946. *The Physical Background of Perception*. Oxford: Clarendon Press.

Barlow, H. B. 1995. The neuron doctrine in perception. In *The Cognitive Neurosciences*, 1st ed., edited by Gazzaniga, M., 415–435. Cambridge, MA: MIT Press.

Barlow, H. B., Kaushal T. B., Hawken M., and Parker A. J. Human contrast discrimination and the contrast discrimination of cortical neurons. *Journal of the Optical Society of America (A)* 4:2366–2371.

Britten, K. H., Shadlen, M. N., Newsome, W. T., and Movshon, J. A. 1992. The analysis of visual motion: A comparison of neuronal and psychophysical performance. *Journal of Neuroscience* 12:4745–4765.

Kreiman, G., Fried, I., and Koch, C. 2002. Single-neuron correlates of subjective vision in the human medial temporal lobe. *Proceedings of the National Academy of Sciences USA* 99:8378–8383.

Rieke, F., Warland, D., van Steveninck, R. R. D., and Posalek, W. 1996. *Spikes: Exploring the Neuronal Code.* Cambridge, MA: MIT Press.

Talbot, W. H., Darian-Smith, I., Kornhuber, H. H., and Mountcastle, V. B. 1968. The sense of flutter-vibration: comparison of the human capacity with response patterns of mechanoreceptive afferents from the monkey hand. *Journal of Neurophysiology* 31:301–334.

Vallbö, X. B. 1989. Single fibre microneurography and sensation. In *Hierarchies in Neurology: A Reappraisal of the Jacksonian Concept,* edited by Kennard, C., and Swash, M. 93–109. London: Springer-Verlag.

Werner, G., and Whitsel, B. L. 1968. Topology of the body representation in somatosensory I of primates. *Journal of Neurophysiology* 31:856–860.

A Last-Minute Change in Worldview

Ramón y Cajal, S. 1995. *Histology of the Nervous System of Men and Vertebrates.* Vol. 2. Translated from the 1911 French ed. New York: Oxford University Press. Cajal's descriptions and interpretations of the crossing-over of retina axons are on pages 304–313.

Retrieving a Lost Dimension/
Information Processing in the Brain from the Bottom Up

Barlow, H., Blakemore, C., and Pettigrew, J. D. 1967. The neuronal mechanisms of binocular depth and discrimination. *Journal of Physiology* 193:327–342.

Berlucchi, G., and Rizzolatti, G. 1968. Binocularly driven neurons in visual cortex of split-chiasm cats. *Science* 159:308–310.

Blakemore, C. 1970. The representation of three-dimensional visual space in the cat's striate cortex. *Journal of Physiology* 209:155–178.

Churchland, P. M. 1992. A feed-forward network for fast stereo vision with a movable fusion plane. In *Proceedings of the 2nd International Workshop on Human and Machine Cognition,* edited by K. Ford and C. Glymour. Cambridge, MA: MIT Press.

Frisby, J. P. 1980. *Seeing: Illusion, Brain and Mind.* Oxford: Oxford University Press.

Hubel, D. H. 1982. Exploration of the primary visual cortex, 1955–1978 (Nobel lecture). *Nature* 299:515–524.

Hubel, D. H., and Wiesel, T. N. 1968. Receptive fields and functional architecture of monkey striate cortex. *Journal of Physiology* 195:215–243.

Johnson-Laird, P. N. 1988. *The Computer and the Mind.* Cambridge, MA: Harvard University Press.

Julesz, B. 1960. Binocular depth perception of computer-generated patterns. *Bell System Technical Journal* 39:1125–1162.

Julesz, B. 1971. *Foundations of Cyclopean Perception.* Chicago: University of Chicago Press.

Longuet-Higgins, H. C. 1981. A computer algorithm for reconstructing a scene from two projections. *Nature* 293:133–135.

Marr, D. 1982. *Vision: A Computational Investigation into the Human Representation and Processing of Visual Information.* San Francisco: W. H. Freeman.

Marr, D., and Poggio, T. 1979. A computational theory of human stereo vision. *Proceedings of the Royal Society of London B* 204:301–328.

Poggio, G. F., and Poggio, T. 1984. The analysis of stereopsis. *Annual Review of Neuroscience* 7:379–412.

Rolls, E. T., and Treves, A. 1998. *Neural Networks and Brain Function.* New York: Oxford University Press.

Schraudolph, N. W., and Sejnowski, T. J. Competitive anti-Hebbian learning of invariants. In *Advances in Neural Information Processing Systems,* edited by Moody, J. E., Hansen, S. J., and Lippmann, R. R., Vol. 4, 1017–1024. San Mateo, CA: Morgan Kaufmannn.

Zeki, S. 1999. *Inner Vision: An Exploration of Art and the Brain.* Oxford: Oxford University Press.

Being of One Mind

Gazzaniga, M. S. 1995. Principles of human brain organization derived from split-brain studies. *Neuron* 14:217–228.

Schiller, F. 1979. *Paul Broca: Founder of French Anthropology, Explorer of the Brain.* Berkeley: University of California Press.

Sperry, R. W. 1964. The great cerebral commissure. *Scientific American* 210:42–52.

Two Minds in One Body?/An Old Pathway Between Brain Hemispheres for Information Producing Emotion

Bogen, J. E. The callosal syndromes. In *Clinical Neuropsychology,* 3rd ed., edited by Heilman, K. M., and Valenstein, E. New York: Oxford University Press.

Mark, V. 1996. Conflicting communicative behavior in a split-brain patient: Support for dual consciousness. In *Towards a Science of Consciousness,* edited by Hameroff, S. R., Kraszniak, A. W., and Scott, A. C., 189–196. Cambridge, MA: MIT Press. (This chapter reports on the interview mentioned in the text.)

Sperry, R. 1974. Lateral specialization in the surgically separated hemispheres. In *Neuroscience 3rd Study Program,* edited by Schmitt, F. O., and Worden, F. G., 5–19. Cambridge, MA: MIT Press.

Sperry, R. 1982. Some effects of disconnecting the cerebral hemispheres (Nobel lecture). In *Les Prix Nobel*. Stockholm: Almquist & Wicksell International.

The Virtues of Parallel Computation

Reichardt, W., and Poggio, T. 1976. Visual control of orientation behaviour in the fly: Part I, A quantitative analysis. *Quarterly Reviews of Biophysics* 9:311–375.
Thorpe, S., Fize, D., and Marlot, C. 1996. Speed of processing in the human visual system. *Nature* 381:520–522.

15. Information Transforms in the Cortex and the Genesis of Meaning

What the Eyes Tell the Brain/Coding for the Vertical, the Horizontal, and the Oblique

Dowling, J. E. 1987. *The Retina: An Approachable Part of the Brain*. Cambridge, MA: Harvard University Press.
Hubel, D. H. 1995. *Eye, Brain, and Vision*. Scientific American Library. New York: W.H. Freeman & Co.
Hubel, D. H., and Wiesel, T. N. 1959. Receptive fields of single neurons in the cat's striate cortex. *Journal of Physiology* 148:574–591.
Kuffler, S. W. 1953. Discharge patterns and functional organization of the mammalian retina. *Journal of Neurophysiology* 16:37–68.

The Grandmother Cell

Barlow, H. B. 1995. The neuron doctrine in perception. In *The Cognitive Neurosciences,* 1st ed., edited by Gazzaniga, M., 415–435. Cambridge, MA: MIT Press.
Gross, C. G. 2002. Genealogy of the "grandmother cell." *The Neuroscientist* 8: 512–518.
Gross, C. G., Bender, D. B., and Rocha Miranda, C. E. 1969. Visual receptive fields of neurons in inferotemporal cortex of the monkey. *Science* 166:1303–1306.
Konorsky, J. 1967. *Integrative Activity of the Brain: An Interdisciplinary Approach*. Chicago: University of Chicago Press.
Kreiman, G. 2001.On the neuronal activity in the human brain during visual recognition, imagery and binocular rivalry. PhD thesis. Pasadena: California Institute of Technology.
Kreiman, G., Fried, I., and Koch, C. 2002. Single neuron correlates of subjective vision in the human medial temporal lobe. *Proceedings of the National Academy of Sciences USA* 99:8378–8383.

Kreiman, G., Koch, C., and Fried, I. 2000. Category-specific visual responses of single neurons in the human medial temporal lobe. *Nature Neuroscience* 3:946–953.

Distributed Coding

Barlow, H. B. 1972. Single units and sensation: a neuron doctrine for perceptual psychology. *Perception* 1:371–394.

Baylis, G. C., Rolls, E. T., and Leonard, C. M. 1995. Selectivity between faces in response to faces of the population of cells in cortex of the superior temporal sulcus of the monkey. *Brain Research* 342:91–102.

Braitenberg, V., and Schütz, A. 1991. *Anatomy of the Cortex.* Heidelberg: Springer-Verlag.

Livingston, M., and Hubel, D. 1988. Segregation of form, color, movement and depth. *Science* 240:740–750.

Rolls, E. T., and Treves, A. 1998. *Neuronal Networks and Brain Function.* Oxford: Oxford University Press.

The Imprinting of Cortical Cells

Logothetis, N. K., and Pauls, J. 1995. Psychophysical and physiological evidence for viewer-centered object representations in the primate. *Cerebral Cortex* 5:270–288.

Logothetis, N. K., Pauls, J., and Poggio, T. 1995. Shape representation in the inferior temporal cortex of monkeys. *Current Biology* 5:552–563.

Logical Depth

Bennett, C. H. 1988. Logical depth and physical complexity. In *The Universal Turing Machine: A Half-Century Survey,* edited by Herken, R., 227–258. New York: Oxford University Press.

Chaitin, G. 1977. Algorithmic information theory. *IBM Journal of Research and Development* 21:350–359.

Kolgomorov, A. N. 1965. Three approaches to quantitative definition of information. *Problems in Information Transmission* 1:1–7.

Lowbrow Meaning

García-Añoveros, J., and Corey, D. P. 1997. The molecules of mechanosensation. *Annual Review of Neuroscience* 20:567–598.

Loewenstein, W. R. 1965. Facets of a transducer process. *Cold Spring Harbor Symposia in Quantum Biology* 30:29–43.

Loewenstein, W. R., and Rathkamp, R. 1958. The sites for mechano-electric conversion in a Pacinian corpuscle. *Journal of General Physiology* 41:1245–1265.

Loewenstein, W. R., and Skalak, R. 1966. Mechanical transmission in a Pacinian corpuscle: An analysis and a theory. *Journal of Physiology* 182:346–378.

Mendelson, M., and Loewenstein, W. R. 1966. Mechanisms of receptor adaptation. *Science* 144:554–555.

Pacini, F. 1835. Sopra un particulare genere di piccoli corpi globulari scoperti nel corpo umano da Filippo Pacini. In *Archivio delle scienze mediche-fisiche* 8.

Sachs, F. 1990. Stretch-sensitive ion channels. *The Neurosciences* 2:49–57.

16. The Conscious Experience

The Conscious Experience of the Passing of Time, and Memory

Crick, F., and Koch, C. 2003. A framework for consciousness. *Nature Neuroscience* 6:119–126.

Sacks, O. 2007. A neurologist's notebook: The abyss. *New Yorker,* September 24, 100–111.

Sperling, G. 1969. The information available in brief visual presentations. *Psychological Monographs* 74:1–30.

Wearing, D. 2005. *Forever Today: A Memoir of Love and Amnesia.* London: Corgi Books.

Unconscious Thinking

Crick, F., and Koch, C. 2000. The unconscious homunculus. *Neuro-psychoanalysis* 2:3–11.

Freud, S. (1923) 1961. The ego and the id. *Standard Edition* 19:1–59. London; Hogarth Press.

Hadamard, J. 1945. *The Mathematician's Mind.* Princeton, NJ: Princeton University Press.

Hadamard, J. 1954. *An Essay on the Psychology of Invention in the Mathematical Field.* New York: Dover Publications.

Jackendoff, R. 1987. *Consciousness and the Computational Mind.* Cambridge, MA: MIT Press.

Poincaré, H. 1913. Mathematical creation. In *Foundations of Science.* New York: The Science Press.

From Piecemeal Information to Unity of Mind

Crick, F., and Koch, C. 2003. A framework for consciousness. *Nature Neuroscience* 6:119–126.

Edelman, G. M. 1989. *The Remembered Present: A Biological Theory of Consciousness.* New York: Basic Books.

Edelman, G. M., and Tononi, G. 2000. *A Universe of Consciousness.* New York: Basic Books.

Hoffmann, H. 1845. *Struwwelpeter.* (The bit about the hare is paraphrased from "Die Geschichte von dem Wilden Jäger" in this old German children's book.)

Koch, C., and Greenfield, S. 2007. How does consciousness happen? *Scientific American* 297(4):76–83.

Llinás, R., Ribary, U., and Wand, X.-J. 1994. Content and context in temporal-thalamo cortical binding. In *Temporal Coding in the Brain,* edited by G. Buszaki, G., et al. Berlin: Springer-Verlag.

Feelings and Emotions/Gut Feelings

Bernard, C. 1878. *Leçons sur les Phénomènes de la Vie Communes aux Animaux et aux Végetaux.* Paris: Baillière.

Craig, A. D. 2002. How do you feel? Interoception: The sense of the physiological condition of the body. *Nature Reviews Neuroscience* 3:655–666.

Damasio, A. 1999. *The Feeling of What Happens: Body and Emotion in the Making of Consciousness.* New York: Harcourt.

Damasio, A. 2003. *Looking for Spinoza; Joy, Sorrow and the Feeling Brain.* New York: Harcourt.

LeDoux, J. 2002. *Synaptic Self.* New York: Simon & Schuster.

Tononi, G. 2012. *Phi: A Voyage from the Brain to the Soul.* New York: Pantheon Books.

The Virtues of Signal Synchrony as a Higher-Order Sensory Code

Freeman, W. J. 1975. *Mass Action in the Nervous System.* New York: Academic Press.

Malsburg, C. von der. (1981) 1994. The correlation theory of brain function. (Max-Planck Institute for Biophysical Chemistry, Heidelberg, Internal Report 81–2). In *Models of Neural Networks,* edited by Domany et al. Berlin: Springer-Verlag.

Malsburg, C. von der. 1999. The what and why of binding: The modeler's perspective. *Neuron* 24:95–104.

Signal Synchrony and Conscious Perception/Neuronal Synchronization during Competition for Access to Consciousness

Engel, A. K., and Singer, W. 2001. Temporal binding and neural correlates of awareness. *Trends in Cognitive Sciences* 5:16–25.

Fries, P., Roelfsena, P. R., Engel, A. K., König, P., and Singer, W. 1997. Synchronization of oscillatory responses in visual cortex correlates with perception in interocular rivalry. *Proceedings of the National Academy of Sciences* 94:12699–12704.

Gray, C. M. 1999. The temporal correlation hypothesis of visual feature integration: Still alive and well. *Neuron* 24:31–47.

Singer, W. 1999. Neuronal synchrony: A versatile code for the definition of relations. *Neuron* 24:49–65.

17. Consciousness and Quantum Information

Boltzmann's World

Cassirer, E. 1956. *Determinism and Indeterminism in Modern Physics.* New Haven, CT: Yale University Press.

Exner, F. 1908. *Über Gesetze in Naturwissenschaft und Humanistik.* Vienna: Selbstverlag der K.K. Universität.

Where the Quantum World May Come into Evolution's Ultimate Information Enterprise

Byron, Lord G. 1823. *Don Juan.* Canto 14.

Quantum Information Waves/The Vanishing Act

Broglie, L. de. 1924. *Recherche sur le Theorie des Quanta: Thèses presentée a la Faculté des Sciences de l'Université de Paris.* Paris: Moisson & Cie.

Dirac, P. A. M. 1929. Quantum mechanics of many-electron systems. *Proceedings of the Royal Society of London* A123:713–733.

Greene, B. 2004. *The Fabric of the Cosmos: Space, Time, and the Texture of Reality.* New York: Alfred Knopf.

Joos, E. 2010. Decoherence through interaction with the environment. In *Decoherence and the Appearance of a Classical World in Quantum Theory,* edited by Joos, E., Zeh, H. D., Kiefer, C., Giuliani, D. G., Kupsch, I., and Stamatesen, I.-O. Berlin/Heidelberg: Springer-Verlag.

Moore, W. 1990. *Schrödinger: Life and Thought.* Cambridge, UK: Cambridge University Press.

Schrödinger, E. 1926. Quantisierung als Eigenwertproblem. *Annals of Physics* 79:361–376.

Schrödinger, E. 1944. *What Is Life?* Cambridge, UK: Cambridge University Press.

Are Quantum Waves and Consciousness Entangled?

Hameroff, S. R., and Penrose, R. 1996. Orchestrated reduction of quantum coherence in brain microtubules. In *Toward a Science of Consciousness: The First Tucson Discussions and Debates,* edited by Hameroff, S. R., Kasniak, A.W., and Scott, A.C. Cambridge, MA: MIT Press.

Heisenberg, W. 1958. The representation of nature in contemporary physics. *Daedalus* 87:95–108.

Penrose, R. 1994. *Shadows of the Mind.* Oxford: Oxford University Press.

Penrose, R. 1996. On gravity's role in quantum state reduction. *General Relativity and Gravitation* 28:581–600.

Wigner, E. P. 1979. *Symmetries and Reflections.* Woodbridge, CT: OxBow Press. (Chapter 13, which is reprinted from an earlier article in 1961.)

Irretrievable Quantum Information Losses: Decoherence

Gell-Mann, M., and Hartle, J. B. 1997. Strong decoherence. In *Quantum Classical Correspondence: The 4th Drexel Symposium on Quantum Nonintegrability,* 3–35. Cambridge, MA: International Press.

Joos, E. 2010. Decoherence through interaction with the environment. In *Decoherence and the Appearance of a Classical World in Quantum Theory,* edited by Joos, E., Zeh, H. D., Kiefer, C., Giuliani, D. G., Kupsch, I., and Stamatesen, I.-O. Berlin/Heidelberg: Springer-Verlag.

Zeh, H. D. 1970. On the interpretation of measurement in quantum theory. *Foundations of Physics* 1:69–76.

Zurek, W. H. 2003. Decoherence, einselection and the quantum origins of the classical. *Physics Today* 75:715–775.

On the Possibility of Quantum Coherence in the Brain

Hagan, S., Hameroff, S.R., and Tuszynski, J. A. 2002. Quantum computation in brain microtubules: Decoherence and biological feasibility. *Physical Review E* 65: 061901–061911.

Hepp, K. 1999. Toward the demolition of a computational quantum brain. In *Proceedings of the Xth Max Born Symposium,* Edited by Blanchard, P., and Jadczyk, A., 92–104. Berlin/Heidelberg: Springer-Verlag.

Tegmark, M. 2000. Importance of quantum coherence in brain processes. *Physical Review E* 61:4194–4206.

18. Molecular Quantum Information Processing and Quantum Computing

Quantum Computers

Benioff, P. 1982. Quantum mechanical models of Turing machines that dissipate no energy. *Physical Review Letters* 48:1581–1585.

Bennett, C. H., and DiVincenzo, D. P. 2000. Information and computation. *Nature* 404:247–255.

Feynman, R. P. 1982. Simulating physics with computers. *International Journal of Theoretical Physics* 21:467–488.

Shor, P. W. 1994. Algorithms for quantum computation: Discrete logarithms and factoring. In *Proceedings of the 35th Annual Symposium on Foundations of Computer Science,* 124–134. Washington, D.C.: IEEE Computer Society Press.

Quantum Computers in the Real: Computing with Atoms

Boozer, A. D., Boca, A., Miller, R., Northrup, T. E., and Kimble, H. J. 2007. Reversible state transfer between light and a single trapped atom. *Physical Review Letters* 98:193601.

Haroche, S., and Raimond, J. M. 1994. Manipulation of non-classical field states by atom interferometry. In *Cavity Quantum Electrodynamics*, edited by Berman, P. R., 123. Boston: Academic Press.

Leibfried, D., Knill, E., Ospelkaus, C., and Wineland, D. J. 2007. Transport quantum logic gates for trapped ions. *Physical Review A* 76:032324.

Lloyd, S. 1993. A potentially realizable quantum computer. *Science* 261:1569–1571.

Lloyd, S. 2006. *Programming the Universe.* New York: Alfred Knopf.

Milburn, G. J. 1998. *The Feynman Processor.* Cambridge, MA: Perseus Books.

Monroe, A., Meekhof, D. M., King, B. E., Itano, W. M., and Wineland, D. J. 1995. Generation of a fundamental quantum logic gate. *Physical Review Letters* 75:4714.

Turchette, Q. A., Hood, C. J., Lange, W., Mabuchi, H., and Kimble, H. J. 1995. Measurement of conditional phase shifts for quantum logic. *Physical Review Letters* 75:4710.

Quantum Computing with Atomic Nuclei/Natural Quantum-Computer Games/Toward Multi-Qubit Computers

Cory, D. G., Fahmy, A. F., and Havel, T. F. 1997. Ensemble quantum computing by NMR spectroscopy. *Proceedings of the National Academy of Sciences USA* 94:1634–1639.

Cory, D. G., Laflamme, R., Knill, E., Viola, L., Havel, T. F., et al. 2000. NMR based quantum information processing: Achievements and prospects. *Fortschritte der Physik* 48:875–901.

Gershenfeld, N. A., and Chuang, I. L. 1997. Bulk spin-resonance quantum computation. *Science* 275:350–356.

Grover, L. K. 1997. Quantum mechanics helps in searching for a needle in a haystack. *Physical Review Letters* 79:325–328.

Lloyd, S. 2008. Quantum information matters. *Science* 319: 1209–1211.

Ramanathan C., Boulant, N., Chen, Z., Cory, D. G., Chuang, I., and Steffen, M. 2004. NMR quantum information processing. *Quantum Information Processing* 3:15–44.

Ramanathan, C., Sinha, S., Baugh, J., Havel, T. F., and Cory, D. G. 2005. Selective coherence transfers in homonuclear dipolar coupled spin systems. *Physical Review A* 71:020303(R).

Vandersypen, L., Steffen, M., Breyta, G., Yannoni, C., Cleve, R., and Chuang, I. L. 2001. Experimental realization of Shor's quantum factoring algorithm using nuclear magnetic resonance. *Nature* 414:883–887.

Wüthrich, K. 1995. *NMR in Structural Biology.* Singapore: World Scientific.

19. Quantum Information Processing and the Brain

Naturally Slow Decoherence

Bloch, F. 1946. Nuclear induction. *Physical Review* 70:460–474.

Two Variations on an Ancient Biological Quantum Theme

Engel, G. S., Calhoun, T. R., Read, E. L., Ahn, T.-K., Mančal, T., Cheng, Y.-C., Blankenship, R. E., and Fleming, G. R. 2007. Evidence for wave-like energy transfer through quantum coherence in photosynthetic systems. *Nature* 446:782–786.

Hecht, S., Schlaer, S., and Pirenne, M. H. 1942. Energy, quantum and vision. *Journal of General Physiology* 25:819–840.

Read, E. L., Schlau-Cohen, G. S., Engel, G. S., Wen, J., Blankenship, R. E., and Fleming, G.R. 2008. Visualization of excitonic structure in the Fenna-Matthews-Olson photosynthetic complex by polarization-dependent two-dimensional electronic spectroscopy. *Biophysics Journal* 95:847–856.

A Final Darwinistic-Physics Inquierendo and a Hypothesis

Loewenstein, W. R. 2000. *The Touchstone of Life: Molecular Information, Cell Communication, and the Foundations of Life.* New York: Oxford University Press.

Acknowledgments

I thank Thomas Kelleher, my editor at Basic Books, for sound counsel. He offered sharp editorial insights and trenchant comments, including the suggestion of the book title. He always insisted on clarity and would read me the riot act whenever I left a scientific term undefined. That some parts still are not easy to read is my fault, not his.

I thank Tisse Takagi, associate editor at Basic Books, for steering the manuscript so efficiently and pleasantly through the editorial and production stages.

I was fortunate to have James Levine as my agent for a number of years. This book owes a great obligation to his skills in shepherding it from inception to publication.

Paul Oberlander at the MIT-Woods Hole Oceanographic Institute turned my rough sketches into professional figures. We have worked together for nearly two decades on various articles and books and had reached that rare point where we would communicate without words. I thank him for his superb illustrations.

I am obliged to Roger Lester for reading an early version of the book. And I thank Kathy Lynch for unfailing secretarial assistance. She would cheerfully decipher my dreadful handwriting and type and retype page after page and version after version of the manuscript.

I feel a deep gratitude to the Marine Biological Laboratory, Woods Hole, for having provided me for many years during the summer months with a quiet environment for work and thought.

I am indebted to many individuals and institutions in the United States and abroad for arranging and hosting the many lectures I gave over the years. My hosts were too many to be named here, but the opportunities they gave me were invaluable, as they provided an early testing ground for many of the ideas in the book. One speaks to learn as much as to teach, and I learned much from so wide an audience.

For generously supporting my own research for several decades, I gratefully acknowledge the National Science Foundation and the National Institutes of Health.

As always, the indescribable debt is to Birgit, who for so long endured the presence of an extra soul in our home and shared with me the birth pangs.

Illustrations and Other Credits

The verse about the neutrino, Chapter 5, is from a poem entitled "Cosmic Gall," by John Updike, in *Telephone Poles and Other Poems* (New York: Knopf, 1959). A. Knopf, a division of Random House, with permission of Random House.

Figure 7.5 is based on an illustration (Figure 3) in M. S. Singer, Analysis of the molecular basis for octanal interactions in the expressed rat 17 olfactory receptor, *Chemical Senses* 25 (2000):155–165, with permission of Oxford University Press

Figure 8.1 is based on illustrations (Figures 2 and 3) in F. R. Salemme, Structure and function of cytochromes c, *Annual Review of Biochemistry* 46 (1977):299–329, with permission of Annual Reviews, Inc.

Figure 14.3 is based on an illustration (Figure 18, Plate 7) in H. Braak, On the striate area of the human isocortex, *Journal of Comparative Neurology* 166 (1976): 341–364, with permission of John Wiley and Sons.

Figure 14.9 is reproduced from Figure 2.4-1 B. Julesz, *Foundations of Cyclopean Perception* (Chicago: University of Chicago Press, 1971), with permission of University of Chicago Press and Alcatel-Lucent USA, Inc.

Figure 15.3 is based on illustrations by G. Kreiman, On the neuronal activity in the human brain during visual recognition, imagery and binocular rivalry (Ph.D. thesis, Pasadena, California Institute of Technology, 2001); and in G. Kreiman, I. Fried, and C. Koch, Single neuron correlates of subjective vision in the human medial temporal lobe, *Proceedings of the National Academy of Sciences USA* 99 (2002):8378–8383, with permission of National Academy of Sciences USA.

Figure 15.4 is based on illustrations (Figure 10) in N. K. Logothetis and J. Pauls, Psychophysical and physiological evidence for viewer-centered object representations in the primate, *Cerebral Cortex* 5 (1995):270–288, with permission of Oxford University Press.

Figure 15.5 *right* is reproduced from Figure 1 in an article of mine with R. Skalak, Mechanical transmission in a Pacinian corpuscle: An analysis and a theory, *Journal of Physiology* 182 (1966):346–378, with permission of John Wiley and Sons.

Figure 16.4 is based on illustrations in A. K. Engel and W. Singer, Temporal binding and neural correlates of awareness, *Trends in Cognitive Sciences* 5 (2001):16–25, with permission of Elsevier, Ltd.

Figures 17.1 and 17.2 are based on illustrations in E. Joos, Decoherence through interaction with the environment, in *Decoherence and the Appearance of a Classical World in Quantum Theory*, edited by E. Joos, H. D. Zeh, C. Kiefer, D. G. Giuliani, I. Kupsch, and I.-O. Stamatesen (Berlin/Heidelberg: Springer-Verlag, 2010), with permission of E. Joos and Springer-Verlag Heidelberg.

Figure 17.3 is reproduced from an illustration in E. Joos, Decoherence through interaction with the environment, in *Decoherence and the Appearance of a Classical World in Quantum Theory*, edited by E. Joos, H. D. Zeh, C. Kiefer, D. G. Giuliani, I. Kupsch, and I.-O. Stamatesen (Berlin/Heidelberg: Springer-Verlag, 2010), with permission of E. Joos and Springer-Verlag Heidelberg.

Index

Abstraction, 165–166, 192–193, 194, 200, 204
Adrian, Lord Edgar, 78, 229, 281
Alcohols, 93 (fig.)
Aldehydes, 93 (fig.), 94
Algorithmic information theory, 206n
Algorithms
  and the feeling of unreality, 161
  evolutionary, 116–118
  machines for running, 137
  predictions using, 160
  for quantum computing, 252n, 262
  See also Mathematics
Alice in Wonderland (Carroll), 5 (fig.), 163, 204
Allegory of the Cave, 156
Ambiguity, 91, 92 (fig.), 98, 124
Amino-acid modules/units
  as dipoles, 63 (fig.), 69, 70, 97
  reshuffled, 38
  of Rhodopsin, 66, 69, 70, 97
  See also Protein entries
Amygdala, 179n, 197, 200 (fig.), 223, 226–227
AND gate, 257, 260
AND operation, 61, 257
Anderson-Higgs asymmetries, 5n
Anger, recognition of, 226, 227
  See also Emotion
Anticipation, 134, 226
  See also Forecognition
Aristotle, 245
Art of Love (Ovid), 215
Asymmetry of time, 4, 5n, 6–8, 19, 46

Atomic nuclei, 17, 19, 58, 59 (fig.), 64, 66, 75, 80n, 83, 101, 157
  quantum computing with, 254–256, 265, 268
Atoms
  discovery of, 151
  energies of, 75, 76, 77
  formation of, 17, 18 (fig.)
  quantum computing with, 250, 252–254, 260, 262
  quantum states of, and evolutionary strategy, 78
ATP (adenosine triphosphate), 30, 32, 46 (fig.), 73 (fig.), 85, 100, 101, 103
Auditory experiences
  conscious states underlying, time and, 3
  and consciousness, 219–220
Auditory system
  and information streamlining, 196
  ordering principles, 172
  and signal synchrony, 230
Augustine of Hippo, Saint, 2
Awareness and consciousness, 216
Axel, Richard, 89

Baer, Karl Ernst von, 197
Banks, Joseph, 104–105
Baranco, Adriano, 258
Barlow, Horace, 199 (box)
Basal forebrain, 223 (fig.), 226
Bathorhodopsin, 64, 66
Batteries, 104–105
Baylor, Denis, 82

Benioff, Paul, 250
Bennett, Charles, 206, 277
Bérnard, Claude, 224
Beryllium ion, 253
Besso, Michele, 13
Beta(β)-carotene, 24–25, 62
Bias of world picture, 143–145, 167
Big Bang, 6, 14, 15–16, 17, 18 (fig.),
    133, 245
Big Crunch, 6
Binding problem, 221–222, 223, 227
Biological discoveries
    transcending the senses in, 151–154
Biological quantum themes, 267–269
Biomolecular complexity
    first random generator of, 107,
        109–113, 114, 116
    generators of, 109–113, 118–124
    genetic history and the random
        generator of, 114–116
    and genuine transmogrification,
        107–109
    and the heuristics of
        transmogrification, 113–114
    second random generator of,
        118–121, 122
Bipolar cells, 68, 73 (fig.), 166 (fig.)
Bits, computing, 108, 142n, 257
    See also Qubits (quantum bits)
Black holes, 5n
Blankenship, Robert, 268
Bloch, Felix, 266
Blood vessels, emotion and, 224–225
Bohr, Niels, 60, 241
Boltzmann, Ludwig, 12–13, 236, 237,
    266, 277
Boltzmann's equation, 12, 277
Boltzmann's world, 235, 236
Boole, George, 61, 257
Borges, Luis, 8
Bottom-up information processing,
    178–179, 184, 222
Bragg, Lawrence, 152
Bragg, William, 152
Brain computer, 142–143
    See also Information processing
Brain evolution, 96, 71–72, 77–78, 79,
    150, 158, 159, 184
    co-option in, 159, 160

Brain size, 135, 150n, 158
Brain stem, 184, 223, 225
Brain topography
    and cell organization, 166–168,
        169 (fig.)
    cortical-cell, 170–173
    of smell information, 88 (fig.),
        89–90, 184
    of the two hemispheres, 180 (fig.),
        181 (fig.)
    See also specific parts of the brain
Brainstem, 171 (fig.)
Broca, Paul, 158n
Broca's area, 158, 171 (fig.), 180, 208,
    220n
Broglie, Prince Louis de, 65 (box)
Bromothiophene, 260, 261, 262
Buck, Linda, 89

Calcium, origin of, 19
Calcium channels, 37, 38, 55n
Calcium fluoride, 263
Calcium ions, 51, 130
Cambridge Physics Laboratory, 151, 152
Carbon, origin of, 17, 19
Carbon chains
    carotene, 25
    odor, 93, 94, 95
    retinal, 63, 64, 69, 268, 269
Carbon nuclei, 262, 263
Carotenes, 23–25, 41, 62, 85, 131
Cell membranes
    cell-to-cell channels in, 222
    channels in, 37, 40, 42, 45 (fig.)
    diagram of, 52
    evolution's use of, 35
    ion channels in, 54–55, 82
    lipids of, 51–52, 53–54, 56
    and the sensory transducer unit, 49
Cell Theory, 68n
Cell water, 35, 42, 43, 44, 55, 82, 104, 111
    See also Cytoplasm
Cell-cell recognition
    genes encoding for, 120–121
    odor demons and, 98
Censoring of information, 46–48,
    189–190, 191, 246
Cerebellum, 171 (fig.), 223 (fig.)
Cerebral cortex. *See* Cortex

Cesium atoms, 253
Chaitin, Gregory, 163n, 206n
Chalmers, D. J., 281
Channel demons, 38–41, 42, 43, 44, 45
  (fig.), 46, 47 (fig.), 83, 127
  and the transducer unit, 49, 50
Changeux, Jean-Pierre, 281
Chemo-electric transducer, 50
Chloride channels, 37, 55n, 146, 147n
Chlorophylls, 23–25, 41, 131, 268
Chromatin, 152
Chromosomes, 152
Chuang, Isaac, 254, 259, 262,
  263 (fig.), 282
Cingulum, 225
Clinton, Bill, 197–200, 202
Coding, 91–92, 167, 260
  distributed, 200–202, 273
  and imprinting, 202–204
  neuron network, synchronicity in,
    228–229
  quantum coding, 253, 259, 260
Cognition
  cycles, 28–32, 37, 46, 48, 147
  molecular, 26–28, 93
  quintessence of, 28–32
Cognition equation, 30, 85, 118, 134,
  147, 277
Cognitive brain deficiencies, 181–183
Cognitive speed, measures of, 185–186
Coherence, quantum-wave, 64, 66, 83,
  239, 268, 274
Coleridge, Samuel, 157
Collapse of the wave function, 241
Color
  nature of, 70–72, 75, 76Color
    spectrum, 58, 59 (fig.), 67
  why we see colors, 67, 70–71, 72,
    75–77
Color blindness, 68
Communications
  cell, 86
  language, 3, 91–92
  one-way versus two-way, 98
  See also Information transmission
Competition, for access to
  consciousness, 231–232, 233 (fig.)
Complex and simple cells in cortex, 193,
  194–196

Computation, parallel. *See* Parallel
  computation
Computation process
  classical versus quantum, 256–258
  of Universal Turing Machines,
    138–140
Computation time
  for gestalt recognition, 229
  as a measure of complexity, 205, 206
  saving, 273
  See also Logical depth
Computer-generated images, 140–142,
  165
Computers
  and consciousness, 217
  operational mode of, versus the
    brain's, 184–185, 186, 187
  protein demons as computers,
    147–148
  See also Universal Turing Machines
Computing operations, 257
Conditional gate, 260, 261
Cones, 68, 72, 73 (fig.), 165, 166 (fig.)
Conscious experience
  feelings and emotions, 223–226
  gut feelings, 226–227
  holistic aspects of, 221–223and
    memory, 216, 218–219
  multicellular involvement in, 159, 247
  neural synchrony and, 227–228
  and the passing of time, 216, 217–218
  unconscious thinking, 219–221
Consciousness, xiii, 2–3, 95n, 184, 216
  and the brain hemispheres, 184
  competition for access to, 231–232,
    233 (fig.)
  as culmination of information
    processing and computing,
    216, 217
  development of, 131–135, 216–217,
    222
  entanglement of quantum waves with,
    242–243, 246, 247, 273
  evolutionary role of, 217
  lack of in mathematical thought, 220
  loss of, 218
  neural feedback involved in, 223and
    neural space, 159
  stream of, and time, 2–4, 219

Consciousness *(continued)*
  subcortical information processing
    leading to, 223–225
  timescales of, physiological, 246
Controlled gate. *See NOT* gate.
Convergent mapping, 92–95, 98
COPY gate, 61, 257
Corpus callosum, 180 (fig.), 223 (fig.)
  formation of, 184
  severing of, 180, 181, 182, 183, 184
Cortex
  cortical-cell imprinting, 202–204
  formation of, 184
  information processing units in the,
    168–170
  and sensory information transforms,
    165, 189–204
  measuring the speed of visual
    information transmission
    to the, 186
  neuron web in the, 167–169
  and the olfactory system, 88 (fig.), 89,
    90 (fig.), 230
  organization of the, 167–168,
    169 (fig.)
  simple and complex cells of, 193,
    194–196
  synaptic stations along the way to the,
    145, 171 (fig.), 172 (fig.)
  topography of the, and worldview,
    170–173
  See also Information processing;
    Information transforms; Visual
    cortex
Cory, David, 254, 263
Cosmological arrow, 6–7
Creation of matter, 16
Crick, Francis, 152, 153n, 222, 236
Cross-over of sensory information lines,
  172 (fig.), 173–175, 181 (fig.)
Cytochrome c, 103, 107, 126

Dalton, John, 68
Damasio, Antonio, 227
Darwin, Charles, xv, 15, 16, 129, 156, 226
Darwinistic physics, 70, 72, 75, 76, 133,
  269, 270
Davies, P., 282
Davy, Humphry, 105

de Broglie, Louis, 65 (box)
Decoherence, quantum-wave, 243–245,
  246, 247, 252, 253, 265, 269
  as information loss, 243–246
  and epistemology, 245, 246
Delbrück, Max, 236
Democritus of Abdera, 151
Demons. *See* Molecular demons
Dennett, D., 281
Dendritic tree, 145–146, 147, 190,
  228, 238
Depth perception, 176–177
Deutsch, David, 140, 142n, 250, 258, 282
Diamond, Jared, 150n
Dibromothiophene, 259–260
Dickerson, R. E., 102 (fig.)
Digital demons, 127, 128
  plants and, 128–130
  rise of, 127–128
Digital electrical signals
  evolutionary beginning of, 35, 36
  generation of, 54–56, 88, 127, 210, 211,
    212n, 213, 278
  synchronization of, 222
Diodes, tunnel, 103
Dipoles, 40 (fig.), 63 (fig.), 69, 70, 97
Dirac, Paul, 153, 241
Disparity, visual, 176–178, 179
Distributed coding, 200–202, 273
Distributive law of coding, 61
Divergent mapping, 92 (fig.), 98
DNA (deoxyribonucleic acid)
  compounded scripts of, 113
  discovery of DNA structure as an
    example of sensory transcendence,
    152–153
  error suppressing loops in
    information transmission to
    protein, 111–112, 113n, 114
  human and mouse scripts, 115
  large-scale rearrangements in, 118–121
  mutations of, 109–113, 115, 117
  number of genes contained in, 91
  number of odor-sensing genes in, 89
Dowling, John E., 281
Dyson, F. J., 282

Earth, formation of, 18 (fig.)
Eckert, Artur, 258

Edelman, Gerald, 223
Einstein, Albert, 2, 4, 13, 60, 62n, 65 (box), 155, 160, 161, 242
Einstein, Elsa, 155
Electrical signals, sensory. *See* Digital electrical signals
Electrons, 62, 64, 75–76, 78, 79, 238, 250, 267, 268, 269
  discovery of, 105, 150–151
  electronic information transmission in cells, 101–103, 127
  formation of, 16
  interactions of, 153–154
  as leptons, 100–101
  spin of, 255
  tunneling of, 125, 126
Emotion, 87, 153, 182, 183, 184, 199, 223–227
Encoding. *See* Coding
Energy fields, overlapping, 78
Engel, A. K., 233 (fig.)
Entropy
  information and, 19–20, 277
  negative, 84–85, 134
  wresting information from, 28–32
Ephemerides, 208
Error-suppressing feedback loops, 111
Evolution
  Boltzmann on, 236
  and consciousness, 216–217
  game plan of, 77, 78
  and genetic fine-tuning, 114, 115, 116
  and information economy, 44, 78, 185, 201, 206–207, 249, 273
  of neural networks, 71–72, 77–78, 79, 150, 158, 185
  and parallel computing, 185
  personification of, xv
  trans-adaptations in, 160
Exons, 119, 120, 124

Face recognition, 196, 197–200, 201, 226
Factorization, 251–252
Fear, 200, 224, 225, 226, 227
Feelings. *See* Emotion
Feynman, Richard, 205, 250, 282
Fleming, Graham, 268
Flight apparatus of house fly, 186
Flight simulators, 141

Fly, brain of a, and information processing, 185–186, 187
Forecognition, 134–135, 149
  and mathematics, 159, 160–161, 162
Franklin, Rosalind, 153
Freud, Sigmund, 220n
Fried, Itzhak, 197
Fröhlich, Herbert, 274
Frontal lobe, 171 (fig.), 182

Galaxies. *See* Stars and galaxies
Galileo, 157
Gamma(γ)-aminobutyric acid (GABA), 146, 147n, 168n
Gamma(γ)-rays, 58, 59 (fig.), 109, 110
Ganglion cells, 68, 73 (fig.), 166 (fig.), 191–192, 210
Gauss, Carl Friedrich, 220
Gazzaniga, M. S., 180
G-demons, 84–86, 87, 88
GDP (guanine diphosphate), 84 (fig.)
Gell-Mann, Murray, 245
Gemeingefühl, 225
Gene reshuffling, 118, 119
Generalization, learning, 204
Generator current, 54 (fig.), 145, 201, 210, 211, 212, 213, 278
Genetic coding, 119, 120, 124
Genetic fine-tuning, 114, 116, 118–119, 120–121
Geniculate nucleus, 166 (fig.), 171, 173 (fig.), 192–193, 195 (fig.)
Gershenfeld, Neil, 254, 259, 262
Gestalt-recognition cells, 196–197, 199–200, 201, 202, 203–204, 229
Gleick, K. J., 282
GMP (guanosine monophosphate), cyclic, 82
Gödel, Kurt, 163n
G-proteins, 84, 85
Grand Climacteric, 18 (fig.), 121, 128, 131–132, 133, 224
"Grandmother cells," 196–197, 198–199 (box), 201, 202, 203 (fig.), 204, 205
Gravitational field, 5n, 17, 155
Gravitational force, 6, 8, 58
Gravitons, 22, 58, 242
Gravity, 17, 18 (fig.), 22, 50, 242, 273
Great Leap Forward, 150n

Greene, B., 282
Gross, Charles, 197
GTP (guanine triphosphate), 84 (fig.), 85
Gut feelings, 226–227

Hadamard gate (*H* gate), 258,
    259 (fig.), 260
Hadamard, Jacques, 220, 258
Hartle, Jim, 245
Hearing. *See Auditory entries*
Heisenberg, Werner, 60, 241
Heitler, Walter, 239
Helmholtz, Hermann von, 68n
Hodgkin, Alan, 237
Hofstadter, D., 282
Holistic mind, 179, 221–223
Homeostasis, 224, 225
Hominid brains, 150n
Hopfield, John, 126
Hubble, Edwin, 155
Hubel, David, 192, 193, 194, 195,
    199 (box)
Huxley, Andrew, 237
Hydrogen nucleus, 17, 18 (fig.), 19,
    101n, 153n
Hypothalamus, 223
Hypothesis of neuronal quantum
    processing and computing,
    270–273

Iconic memory, 218
Ikats, 122–124
Imprinting of cortical cells, 202–204
Incompleteness theorem, 163n
Inferior temporal cortex, 171 (fig.), 179n,
    196, 201, 203
Information
    begetting order, 25
    density of, in ordinary language
        versus mathematics, 157–158
    and entropy, 19–20, 277
    evolutionary selection of, 114
    wresting, from entropy, 28–32
    See also specific types of information
Information arrows
    and the arrow of time, 19, 21
    described, 19–21, 96, 266, 267
    of life, 22–25, 32–33, 62
    molecular demons and, 26–28

products of, overview of, 21–22
    See also Second Information Arrows
Information censorship, 46–48, 189–190,
    191, 246
Information loops, 25–26, 111, 114, 133,
    142, 227
Information economy, 46, 78, 103, 128,
    185, 201, 206–207, 227, 249, 269,
    270, 273
Information loss
    quantum, irretrievable, 243–245
    selective, meaning from, 204, 205
Information processing in brain, 237,
    247
    and autonomy of the hemispheres,
        181, 182, 183
    bottom-up, 178–179, 194, 222
    and cell organization in the brain,
        166–168
    and change in worldview from partial
        cross-over of sensory information
        lines, 173–175
    consciousness as a culmination of,
        and computing, 216–217, 222
    and cortical information-processing
        units, 168–170
    and cortical-cell topography and
        worldview, 170–173
    evolutionary role of, 217
    parallel processing and computation,
        184–187
    speed of, 185–186
    and retrieving the third dimension
        (depth), 175–178
    top-down, 179
    See also Quantum information
        processing
Information theory, 96, 108n, 205, 206n,
    230, 262
Information transforms, 114, 141, 143,
    145, 165, 189–214, 212, 278
Information transmission, 237
    from DNA to protein, 111–112, 113n,
        114, 122, 124
    electronic, 99–105, 125–126, 127
    fast, demons for, 36–41
    mechanical, in Pacinian corpuscle,
        279
    by molecular demons, 26, 27

quantum, coherent, 64–66, 83, 239, 268, 274
quantum into molecular, 81–86
sensory, generalized scheme, 42–44
speed of, measuring, 186
Informational continuity, 22, 33
Infrared rays, 22, 50, 59 (fig.), 74, 77, 144
Infrared spectroscopy, 274
Inhibitory and excitatory inputs, 145–146, 168n, 190
Insula, 225, 226
International Nuclear Test-ban Treaty, 138
Interoception, 225–226
Interoceptors, 225
Introns, 119 (fig.), 120
Ion channels, 41, 54–55, 82, 127, 128, 210, 238, 273
  See also Calcium channels; Chloride channels; Potassium channels; Sodium channels
Ion traps, 253, 254
Iron, 19

Jackendorff, R., 281
Johns Hopkins University's Wilmer Institute, 191
Joos, Erich, 245
Julesž, Béla, 177n

Kandel, Eric R., 281
Kant, Immanuel, 149–154, 194
Kaufmann, Walter, 151
Kimble, Jeff, 253, 254
Klangfarbe, 71
Knowledge
  limits of, 162–163
Koch, Christof, 146, 197, 222, 281
Kolmogorov, Andrei, 206n
Konorski, Jerzi, 199 (box)
Kreiman, Gabriel, 197
Kuffler, Stephen, 191–192

Langevin, Paul, 65 (box)
Language center, 181 (fig.)
  See also Broca's area
Language communications, 3, 91–92, 95n, 157–159, 219–220

Left brain, 172 (fig.), 173, 180 (fig.), 181 (fig.), 182
Leptons, 99–101
Lettvin, Jerome, 198 (box)
Llinás, Rodolfo, 223, 303
Lloyd, Seth, 253, 259, 282
Loewenstein, Werner, 54 (fig.), 92 (fig.), 209 (fig.), 210, 211 (fig.), 274, 278, 279
Logic
  laws of, 61, 66
  operators of, 61
  quantum versus ordinary, 60–61, 257
  See also Quantum logic
Logic gates, 257, 258, 259, 260, 262, 265, 268, 274
Logic operations, 61, 253–254, 257, 258, 260, 271
Logic-switch speed/rate, 185, 186, 229
Logical depth, 206–208, 273, 277
Logothetis, Nikos, 202
London, Fritz, 239
Long-term memory, 219

MacKinnon, Roderick, 39
Macromolecules
  information cycles and, 25–26
  search for quantum-logic gates in, 274
  See also Molecular demons
Magnetic resonance imaging (MRI), 254
  See also Nuclear-magnetic-resonance spectroscopy
Malsburg, Christoph von der, 228–229
Mao Tse Tung, 275
Mark, Victor, 183
Mathematics
  and consciousness, 159
  density of information in, 157–158
  evolutionary considerations, 159–160
  and forecognition, 159, 160–161, 162
  limitations of, 162–163
  and reality, 156–157, 160–161, 162
Mathies, Richard, 64, 66
Maxwell, James, 27–28, 68n
Maxwellian demons. *See* Molecular demons
Meaning
  begetting of, 204
  and complexity, 206

Meaning *(continued)*
  ecumenecity of, 205
  logical depth, 206–208
  and information economy, 207
  measuring, 205–206
  progressive increase in brain, 195–196
  Pacinian corpuscle contributing to, 213
Mechano-electric transducers, 50, 51
Medial temporal cortex, 197, 200 (fig.)
Medulla, 172 (fig.)
Membrane channels. *See* ion channels
Memory
  and autonomy of brain hemispheres, 183
  and the conscious experience, 216, 218–219
Mendel, Gregor, 151–152
Metal proteins, 101, 102 (fig.), 103–105, 125, 126, 127
Microwaves, 59 (fig.)
Milky Way, 15, 19
Miller, George, 219
Molecular demons, 26–28, 32
  emergence of, 33
  for fast information transmission, 36–41
  See also specific type of molecular demon
Molecular information, from quantum information into, 81–86
Molecular quantum information processing, and quantum computing, 249–263, 265–266, 267–269, 270, 274
Molecular sensing
  and the line from nose to cortex, 87–89
  and mapping, coding, and synonymity, 90–92
  numerous odor-information channels in, 89–90
  and quantum synonymity, 97–98
  and sensory synonymity, 92–95, 95–96, 98
Molecules and molecular systems, formation of, 18 (fig.), 236
  See also Atomic nuclei; Atoms

Monod, Jaques, 236
Monroe, Christopher, 253
Mountcastle, Vernon, 28
Multicelled organisms, formation of, 18 (fig.), 121, 127
Muons, 79, 100, 255
Mutations, 109–113, 115, 119 (fig.), 120

Neanderthals, 150n
Necker cube, 231
Necker, Louis, 231n
Negative entropy, 84–85, 134
Neumann, John von, 241
Neuron clusters (glomeruli), 88 (fig.), 89, 90 (fig.), 184
Neuron development, 133–134
Neuron network coding, 228–229
Neuron trellis, 133–134, 135, 142, 149, 150, 159, 224
Neuron web, 36, 142, 158, 167–168, 172, 190, 202, 207, 222, 224, 270, 272
Neuronal computing, 145–148, 185
  See also Information processing
Neuronal quantum computing, 270, 271–273
Neuronal virtual-reality generator, 142–143
Neurons
  evolutionary debut of, 131
  and gestalt recognition, 196–197, 200
  increasingly loaded with meaning, 195–196, 204, 205
  logic-switch rate of, 185, 186, 229
Neutrinos, 16, 79–80, 99, 100, 157, 245, 255
Neutrons, 17, 80n, 262
Newton, Isaac, 2, 4, 9–10, 12–13, 67, 156, 160, 161, 207, 208
Nicholls, John, 281
Nitrogen, 94 (fig.), 95, 129, 153n
Noise, 66, 81, 82, 83, 274
NOT gate, 257, 258, 260, 261
NOT operation, 61, 257
Nuclear-magnetic-resonance spectroscopy, 254, 255, 259, 260, 262, 263, 265, 266, 274
Nuclear resonance frequency, 255
Nucleus, atomic. *See* Atomic nuclei

Occipital lobe, 171 (fig.)
Octanal carbon chain, 93 (fig.), 94
Odor demons
  and cell-cell recognition, 98
  described, 87–89
  evolution of, 96, 97
  mapping, 94, 95
  See also Olfactory system
Odor molecules, 87, 93–95
Olfactory experiences
  and emotion, 184
  split-brain effect on, 184
Olfactory system, 88 (fig.)
  and brain topography of smell
    information, 89–90, 184
  gene inactivation in evolution,
    272–273n
  information mapping, 91, 92 (fig.),
    93–95
  sensory scheme involving the, 43
  and signal synchrony, 229–230
  See also Molecular sensing
Optic nerve, 166 (fig.)
OR operation, 61
Orbifrontal cortex, 179n
Orbitals, 76, 77
Orbits, 58, 60, 69, 76
Origin of life, 16, 17, 18 (fig.)
Origin of matter, 16–17, 18 (fig.)
Ovid, 215
Oxygen
  formation of, 17, 19

Pacini, Filippo, 208–209
Pacinian corpuscle, 54, 144, 172,
    208–213, 225, 278, 279
  pressure field in, 210–212, 279
Pain, 172, 173, 226
Parallel processing and computation
  in brain, 184–187, 201–202,
    249, 251
  and distributed coding, 201
  See also Quantum computing,
    quantum information processing
Parietal cortex, 171 (fig.), 172
Paul, John, 202
Penrose, Roger, 242, 243n, 247
Phaedo (Plato), 156

Phase-space trajectory, 116, 117
Philosophy, lesson for, 245–246
Phonological images, 219–220, 221
Photocells, 62
Photo-electric effect, 62n
Photo-electric transducers, 50, 51, 62
Photoelectricity, 64
Photon information, 68, 71, 73 (fig.),
    82, 83
Photon spectrum, 58, 59 (fig.)
  visible, 59 (fig.), 67, 76
Photon traps
  and computing with atoms,
    253, 254
  molecular stability and, 67
  in photosynthesis, 23, 24, 25, 33
  in vision, 62–63, 64
Photon-induced mutations, 109–111,
    113n
Photons, 22, 99, 238, 250, 268, 269
  energies of, 21, 75, 76, 77, 78
  and quantum sensing, 58, 59, 60,
    62, 80
  spin, 101n, 255
  virtual, 60
Photosynthesis, 267–268, 270
  quantum coherence in, 268, 269
Planck's equation, 65 (box), 266
Plato, 156
Poggio, Tomaso, 202
Poincaré, Henri, 13, 14, 116, 220
Polarization
  effect on nuclear spin, 275
Polya, George, 220
Potassium channels, 37, 38, 39–41, 45
    (fig.), 55n, 147n, 210n
Potassium ions, 54–55, 238
Primary sensory cortex, 167,
    169 (fig.), 179
Protein demons
  cognitive cycle of, 28–32, 37, 147
  and DNA mutations, 111
  information states of, 147–148
  of sensory cells, 49, 56
  vision and, 57, 67
  See also specific types of protein demon
Protein dipoles, 69, 70
Protein transfigurations, 108

Protein transmogrification
  heuristics of, 113–114
  mechanisms of, 109–113
  overview of, 107–109
Proteins, 27, 84, 100, 107, 115n
  in cell-cell recognition, 120–121
  common genetic traits of, 109
  metal, 101, 102 (fig.), 103–105, 125,
    126, 127
  as molecular demons, 27, 117
Protons, 16–17
Proust, Marcel, 87, 219
Pyramidal cells, 168, 169–170, 222, 227,
  228, 229
Pythagoras, 156

Quantum bits. *See* Qubits (quantum
  bits)
Quantum coherence, 64, 66, 83, 239,
  268, 274
Quantum computers, definition of, 250
  multi-qubit, 261–263
Quantum computing
  advantages of, 249, 251–252, 273
  with atomic nuclei, 254–256, 265, 268
  with atoms, 250, 252–254, 260, 262
  versus digital computing, 141–142
  and quantum-logic gates, 259–260
  molecular quantum information
    processing and, 249–263, 265–266,
    267–269, 270, 274
  neuronal, 270, 271–273
  two-qubit model of, 259–260
Quantum electron tunneling, 125–126
Quantum gates, 258, 259–260, 268, 274
Quantum gravity, 18 (fig.), 273
Quantum information boosting, 81–83
Quantum information processing and
    quantum computing, 249–263,
    265–266, 267–269, 270, 274
  and the brain, a hypothesis, 270, 273
  neuronal quantum computing, 66,
    270, 271–273
  in photosynthesis, 268, 270
Quantum logic, 60, 61, 66, 257, 258, 259,
  260, 261, 265
Quantum random generator of
  molecular form, 107, 109–113, 114,
  116, 118, 122

Quantum sensing
  in color vision, 67, 70–71, 72, 75,
    76–77
  and coherence, 64–66
  evolutionary strategy of, 77–78, 97
  our windows to the quantum world,
    61–66
  overview of, 57–58
Quantum states
  robust, 265
  superconductive, 268
  superposed, 242–243, 250, 252,
    253, 255, 256 (fig.), 257, 258, 259,
    268, 271
  three-bit gate, 258
Quantum synonymity, 97–98
Quantum theory, 5n, 60, 100, 153, 162,
  241, 245, 250
Quantum waves
  coherence of, 64, 65 (box), 66, 83, 239,
    268, 274
  collapse of the wave function,
    240–241
  decoherence of, 243–245, 246, 247,
    252, 253, 265, 269
  entanglement of, with consciousness,
    242–243, 246, 247, 273
  as information waves, 75–76, 83,
    239, 243
  and neuronal quantum computing,
    270, 271
Quantum world, 58–61
Quantum-logic operations, 61,
  253–254
Quarks, 16, 17, 65 (box), 101, 157,
  255
Qubits (quantum bits), 142n, 239

Radio waves, 58, 59 (fig.)
Ramón y Cajal, Santiago, 166, 220n
Random generators. *See* Quantum
  random generator; Second random
  generator
Reality
  expanded, 155–162, 163
  mathematics and, 156–157,
    160–161, 162
  perceptions of, virtual-reality
    generators and, 143

Receptors. *See* sensory demons and
    transducers
Reichardt, Werner, 186
Relativity theory, 2, 4, 5n, 153, 155,
    157, 242
Remembrances of Things Past (Proust),
    219
Rendering the world
    by computer, 137, 140–142
    by neurons, 142–143, 145–148
    correcting our bias in, 148–149
    and transcending the sensory sphere,
        149–154
    See also Information processing;
        Information transforms
Republic (Plato), 156
Retina, 73 (fig.), 165 (fig.)
Retinal, 62–63, 64, 66, 69, 72, 83,
    268, 269
Retrodiction, 163
Rhodopsin, 63, 66, 67, 68, 69, 70–72,
    73 (fig.), 81–82, 83, 84 (fig.), 85, 86,
    97, 119, 238, 268, 269
Rhodopsin activation, 66–67, 83n
RNA (ribonucleic acid), 27, 32, 33,
    111–112, 120
    and DNA to protein information
        transmission, 111–112, 113n, 114
    as a molecular demon, 117
Rods, 82, 83n, 165, 166 (fig.)
Rolls, Edmund, 201
Rotation rate of nuclei. *See* Spin rate
Roux, Wilhelm, 152

Sacks, Oliver, 218, 281
Salemme, F. R., 102 (fig.)
Schleiden, Matthias, 68n
Schrödinger, Erwin, 4, 65 (box), 75, 239,
    240, 241, 250
Searle, J. R., 281
Second information arrow
    beginnings of, 35–36, 131
    and consciousness, 131–135
    sensory demon tandem and,
        44–46
    and the demons for fast information
        transmission, 36–41
Second random generator, 118–121, 122
Seeing. *See Vision and Visual entries*

Sensing. *See* Molecular sensing;
    Quantum sensing
Sense of time, 1
Sensory cells, 49, 81, 82, 87, 89, 90 (fig.),
    144, 165, 268
Sensory classification, 49–50
Sensory coding, 91–92
Sensory demons, 41–42, 87
    cognitive act of, 93
    genetic encoding, 120, 121 (fig.)
    and transducers, 62
    See also Odor demons; Vision
        demons; Interoceptors
Sensory receptor adaptation, 47 (fig.), 48
Sensory scheme, generalized, 42–44,
    56, 81
Sequential computing, 184, 185, 186, 187
SETI project, 216
Shank, Charles, 64, 66
Shannon's equation, 108, 195, 205, 245,
    266, 277
Shor, Peter, 252n, 262
Short-term memory, 218–219
Signal coherence, 227–228
Signal correction, 148–149
Signal synchrony, 228, 229
    and conscious experience, 229–231
    and neuronal competition for access
        to consciousness, 231–232,
        233 (fig.)
Signals, digital. *See* Digital electrical
    signals
Simple cortical cells, 193, 194–196
Singer, M. S., 94 (fig.)
Singer, Wolf, 231, 232
Single-cell organisms, formation of,
    and transition from, 18 (fig.), 121,
    127, 131
Skalak, Richard, 209 (fig.), 279
Smell. *See Odor and Olfactory entries*
Smolin, L., 282
Sodium channels, 37, 38, 39, 55n, 210n
Sodium ions, 41, 51
Solomonoff-Kolmogoroff-Chaitin
    Complexity, 108n
Somato-sensory cortex, 171 (fig.), 172,
    179n, 181 (fig.), 196
Spacetime, 4, 5n, 242
Spatial coding, 167

Spatial orientation, abstracting the attributes of, 192–193, 194, 195 (fig.)

Sperry, Roger, 180–181

Spin, 101n, 255

Spinal cord, 171 (fig.), 172 (fig.), 210

Split brain, 180–182, 183, 184

String theories, 13, 157

Superconductivity, 268, 274

Superposition. *See* Quantum states

Swanson, R., 102 (fig.)

Synchronization of signals. *See* Signal synchrony

Synonymity
    and ambiguity, 91, 98
    language, 91–92, 95n
    mapping sensory, 92–95
    quantum, 97–98
    sensory, 95–96

Taste. *See Odor and Olfactory entries*

Tauons, 79, 100, 255

Temporal lobe, 171 (fig.), 179n, 182, 196, 201, 203, 226

Thalamus, 172, 223

Thermal noise, 66, 83n, 274

Thermodynamics arrow, 6, 7–8, 11–12, 19

Thinking unconsciously, 219–221

Thompson, Joseph, 105, 151

Thought, laws of, 61, 259

Tide tables, 207

Time
    as arrow, 1, 19, 21, 217
    asymmetries of, 4, 5n, 6–8, 11, 19
    awareness of, 1–2, 217–218
    and the cosmological arrow, 6–7
    evolutionary niche in the structure of, 133–134, 224
    inner, versus physics time, 4–5
    and our perception of reality, 13–14
    and probabilities, 10–11, 11–12, 13
    reversibility of, 12–13, 14, 268
    stream of consciousness and, 2–4, 219
    structure of, 132, 133, 134
    symmetry of, 4–5
    and the thermodynamics arrow, 6, 7–8, 11–12, 19
    uncoupling of sensory tandems and perceived time, 47 (fig.)

Time spans, evolutionary, 15, 18 (fig.), 33, 35

Tononi, Giulio, 303

Top-down processing, 179

Transducers, 49–51

Transmogrification. *See* Protein transmogrification

Trellis. *See* Neuron trellis

Tunnel diodes, 103

Tunneling, electron, 125–126

Turing, Alan, 137, 138, 140, 250

Turing machines. *See* Universal Turing Machines

Twain, Mark, 107

Ultraviolet rays, 22, 50, 59 (fig.), 109–110, 111

Unconscious thinking, 219–221

Unification theories, 13, 156

Universal computers, 108, 140, 252, 257

Universal Turing Machines, 137–140, 142n, 143, 206, 250

Universe
    developmental time chart of, 17, 18 (fig.)
    expansion of the, 6–7, 17, 20 (fig.), 21, 155
    initial information state of the, 20, 21, 23 (fig.)
    organizational units, 21
    origin, 19
    See also Big Bang

Unwin, Nigel, 37 (fig.)

Updike, John, 79

Urbach-Wiethe disease, 227

van der Waals interactions, 26, 39, 43

Vandersypen, L., 263 (fig.)

Venus flytrap, 129–130

Virtual computer images, 140–141, 165, 252

Virtual photons, 60

Virtual-reality generators
    computer, 140–141, 165
    neuronal, 142–143, 165

Vision, 166, 171 (fig.), 173–178
Vision demons, 57–58, 71
  photon trap of, 62–63
  and quantum computing, 64–66,
    268–270
  and the quantum world, 61–66
  stability of, 67
Visual cortex
  abstraction of stimulus attributes by
    cortical cells, 192–196
  cell types inhabiting the, 169 (fig.)
  and gestalt recognition, 197
  information transforms in, 191–193,
    194
  processing of disparity in the,
    177–179
  and signal synchrony, 230
  split-brain effect on the, 181 (fig.),
    182, 183, 184
  topography of the, 170–171, 173–175
Visual disparity, 176–178, 179
Visual field, 181 (fig.), 182, 184
Visual reaction time, measuring,
  185–186
Visual recognition, 196, 197, 200, 201,
  226–227
  See also Gestalt-recognition cells
Visual system
  ordering principles of the, 170–172
  and the general sensory scheme,
    43, 44
  See also Quantum sensing

Volta, Alessandro, 104, 105n
  his battery, 104n
Voltage sensors. *See* Sensors

Watson, James D., 152, 153n, 236
Wave function, 65 (box), 75–76
  collapse of the, 241, 242, 243, 245, 246
  See also Quantum waves
Waves, quantum. *See* Quantum waves
Wearing, Clive, 218, 219
Wernicke's area, 182
White color perception, 72–75
Wiener, Norbert, 220
Wiesel, Torsten, 192, 193, 194, 199 (box)
Wigner, Eugene, 241, 242, 245
Wineland, David, 253, 254
Worldview
  cortical-cell topography and, 170–173
  distortion, 148
  evolutionary change in, 174,
    174n, 175
Wüthrich, Kurt, 282

Xeroderma pigmentosum disease, 111n
X-rays
  and mutations, 109, 110
  spectrum, 59 (fig.)

Young, Thomas, 67–68

Zeh, Dieter, 243, 245
Zurek, Wojciech, 245